Lecture Notes in Mathematics

Edited by A. Dold and B. Eckmann

752

Michael Barr

*-Autonomous Categories

With an Appendix by Po-Hsiang Chu

Springer-Verlag
Berlin Heidelberg New York 1979

Author

Michael Barr
Department of Mathematics
McGill University
805 Rue Sherbrooke Ouest
Montreal, P.Q./Canada H3A 2K6

AMS Subject Classifications (1980): 18A35, 18B15, 18B30, 22B99, 46A12, 46A20, 46M05, 46M10, 46M15, 46M40

ISBN 3-540-09563-2 Springer-Verlag Berlin Heidelberg New York
ISBN 0-387-09563-2 Springer-Verlag New York Heidelberg Berlin

Library of Congress Cataloging in Publication Data
Barr, Michael.
*-Autonomous categories.
(Lecture notes in mathematics ; 752)
Includes bibliographies and indexes.
1. Categories (Mathematics) I. Title. II. Series: Lecture notes in mathematics (Berlin) ; 752.
QA3.L28 no. 752 [QA169] 510'.8s [512'.55] 79-21746
ISBN 0-0-387-09563-2

Printing and binding: Beltz Offsetdruck, Hemsbach/Bergstr.
2141/3140-543210

PREFACE

The category of finite dimensional vector spaces over a field K has many interesting properties: It is a symmetric closed monoidal (hereafter known as autonomous) category which has an object K, with the property that the functor (-,K), internal Hom into K, induces an equivalence with its opposite category. Similar remarks apply to the category of finite dimensional (real or complex) banach spaces. We call such a category *-autonomous. Almost the same thing happens with finite abelian groups, except the "dualizing object", $\mathbb{R}/\mathbb{Z}$ or $\mathbb{Q}/\mathbb{Z}$, is not an object of the category. In no case is the category involved complete, nor is there an obvious way of extending both the closed structure and the duality to any of the completions. In studying these phenomena, I came on a fairly general construction which allows you to begin with one of the above categories (and some similar ones) to embed it fully into a complete and cocomplete category which admits an autonomous structure and which, using the original dualizing object, is *-autonomous.

In an appendix, my student Po-Hsiang Chu describes a construction which embeds any autonomous category into a *-autonomous category. The embedding described is not, however, full and is completely formal.

The work described here was carried out during a sabbatical leave from McGill University, academic year 1975-76 mostly at the Forschungsinstitut für Mathematik der Eidgenossische Technische Hochschule, Zürich. For shorter periods I was at Universitetet i Aarhus as well as l'Universite Catholique de Louvain (Louvain-la-Neuve) and I would like to thank all these institutions. I was partially supported during that year by a leave fellowship from the Canada Council and received research grants from the National Research Council and the Ministère de l'Education du Québec.

Preliminary versions of part of this material has been published in the five papers by me listed in the bibliography. The current version was presented in a series of lectures at McGill in the Winter Term, 1976.

TABLE OF CONTENTS

APPENDIX

CONSTRUCTING *-AUTONOMOUS CATEGORIES

Po-Hsiang Chu

CHAPTER I. PRELIMINARIES

1. Symmetric Closed Monoidal Categories.

(1.1) A symmetric closed monoidal category $\underline{V}$ consists of the following data.

i) A category $\underline{V}_0$;

ii) A functor (tensor) $-\otimes- : \underline{V}_0 \times \underline{V}_0 \to \underline{V}_0$;

iii) A functor (internal hom) $\underline{V}(-,-) : \underline{V}_0^{op} \times \underline{V}_0 \to \underline{V}_0$;

iv) An object I of $\underline{V}_0$;

v) Natural Equivalences

$$r = rV : V\otimes I \to V$$
$$\ell = \ell V : I\otimes V \to V$$
$$i = iV : (I,V) \to V ;$$

vi) Natural Equivalences

$$a = a(V,V',V'') : (V\otimes V')\otimes V'' \to V\otimes(V'\otimes V'')$$
$$p = p(V,V',V'') : (V\otimes V',V'') \to (V,(V',V'')) ;$$

vii) Natural equivalences

$$s = s(V,V') : V\otimes V' \to V'\otimes V$$
$$t = t(V,V',V'') : (V,(V',V'')) \to (V',(V,V'')) ;$$

viii) A natural transformation

$$j = jV : I \to (V,V) ;$$

ix) Natural transformations

$$c = c(V,V',V'') : (V',V'')\otimes(V,V') \to (V,V'')$$
$$d = d(V,V',V'') : (V',V'') \to ((V,V'),(V,V''))$$
$$e = e(V,V',V'') : (V,V') \to ((V',V''),(V,V'')) .$$

(1.2) These data are subject to a great many axioms. The most important of these requires that $\mathrm{Hom}(V,V') \cong \mathrm{Hom}(I,(V,V')$ in such a way that jV corresponds to the identity on V and that the diagram

$$\begin{array}{ccc}
\mathrm{Hom}(I,(V',V''))\times \mathrm{Hom}(I,(V,V')) \cong \mathrm{Hom}(V',V'') \times \mathrm{Hom}(V,V') & & \\
\downarrow & & \downarrow \\
 & & \mathrm{Hom}(V,V'') \\
 & & \downarrow \cong \\
\mathrm{Hom}(I\otimes I,(V',V'')\otimes(V,V')) & \longrightarrow & \mathrm{Hom}(I,(V,V''))
\end{array}$$

commutes. Here the left hand arrow is just an instance of the functoriality of $\otimes$, the bottom arrow uses rI (which, one of the coherence rules states, $= \ell I$) and internal composition while the upper arrow on the right is external composition.

The remaining axioms are of two kinds. Some express the fact there is a great deal of redundancy among the data. For example $\langle p\rangle$ is an isomorphism

$Hom(V\otimes V',V") \to Hom(V,(V',V"))$ and one law says that

$\langle p((V',V"), (V,V'), (V, V"))\rangle \circ (c(V,V',V")) = d(V,V',V")$. Another expresses d similarly by means of p, c and s . Another expresses a in terms of p :

$((V\otimes V')\otimes V",-) \cong (V\otimes V'(V",-)) \cong (V,(V',(V",-))) \cong (V,(V'\otimes V",-)) \cong (V\otimes(V\otimes V"),-)$.

The other kind of axioms are the coherence axioms examplified by MacLane's famous pentagonal axiom which expresses the commutativity of

$$\begin{array}{ccccc} V\otimes(V'\otimes(V"\otimes V"')) & \to & (V\otimes V')\otimes(V"\otimes V"') & \to & ((V\otimes V')\otimes V")\otimes V"' \\ \searrow & & & & \nearrow \\ & V\otimes((V'\otimes V")\otimes V"') & \to & (V\otimes(V'\otimes V")\otimes V"' & \end{array}$$

(1.3) A category $\underline{A}$ is enriched over $\underline{V}$ if there is a functor $\underline{V}(-,-):\underline{A}^{op}\times\underline{A} \to \underline{V}$ such that $Hom(I,\underline{V}(A,B))$ is naturally equivalent to $Hom(A,B)$. Also required is a composition map

$$\underline{V}(B,C)\otimes\underline{V}(A,B) \to \underline{V}(A,C)$$

lying above the composition of morphisms in $\underline{A}$ as well as many coherence axioms which are tabulated in I.5 of [Eilenberg, Kelly]. A tensor

$$-\otimes-:\underline{V}\times\underline{A} \to \underline{A}$$

is a functor such that $-\otimes A:\underline{V} \to \underline{A}$ is, for each $A\in\underline{A}$, left adjoint to $\underline{V}(A,-):\underline{A} \to \underline{V}$. Provided $\underline{A}$ has, and $\underline{V}(A,-)$ commutes with limits, this can usually be shown to exist by the adjoint functor theorem. Analogously it frequently happens that $\underline{V}(-,A):\underline{A}^{op}\to \underline{V}$ has a right adjoint, denoted $[-,A]:\underline{V}^{op} \to \underline{A}$ (note the switch in variance) which determines a bifunctor

$$[-,A]:\underline{V}^{op} \times \underline{A} \to \underline{A}$$

called a cotensor. We have

$$Hom(V,\underline{V}(A,B)) \cong Hom(V\otimes A,B) \cong Hom(A,[V,B]) .$$

(1.4) Let S be a set and n be a cardinal number. An n-ary operation on S is a function $S^n \to S$. If $\Omega = \{\Omega_n\}$ is a class graded by the cardinals n of sets Ω_n a <u>model of</u> or <u>algebra</u> for Ω is a set S equipped with an n-ary operation $\omega S:S^n \to S$ for each $\omega\in\Omega_n$. A morphism $f:S \to T$ of models of Ω is a function such that for all n , all $\omega\in\Omega_n$, the square

$$\begin{array}{ccc} S^n & \xrightarrow{\omega S} & S \\ f^n \downarrow & & \downarrow f \\ T^n & \xrightarrow{\omega T} & T \end{array}$$

commutes.

(1.5) among the elements in Ω_n are assumed to be certain π_i , $i\in n$ where realization in any algebra is the projection to the ith coordinate. If (ω_i), $i \in n$ is an m_i-ary operation and ω is an n-ary operation, then there is an m-ary operation, $m = \Sigma m_i$, whose value on any algebra is the composite

$$S^m \cong \Pi_i S^{m_i} \xrightarrow{(\omega_i)} S^n \xrightarrow{\omega} S .$$

(1.6) The category of algebras and morphisms is called a <u>variety</u>. This notion can most readily be formalized by building a category whose objects are cardinals, whose maps include all the functions and such that

$$\mathrm{Hom}(m,m) = \Omega_n{}^m .$$

In particular m is the sum of that many copies of 1. An algebra is a contravariant product preserving set valued functor and a morphism of algebras is a natural transformation between such functors. Details may be found in [Lawvere] and [Linton] .

(1.7) A full subcategory of a variety which is closed under subalgebras, products and quotients is again a variety. The category of a torsion free abelian groups is an example of a non-varietal full subcategory closed under subalgebras and products but not quotients. We define a <u>quasi-variety</u> to be a full subcategory of a variety closed under inverse limits. Notice that we are not requiring merely that it have these limits but that they be the limits in the variety. We define a <u>semi-variety</u> to be a full subcategory of a variety closed under products and equalizers and which contains the free algebras in the variety.

(1.8) If $\underline{V}$ is a quasi-variety in a variety $\underline{W}$ we may consider $\underline{W}'$ the closure of $\underline{V}$ under quotients. It is a standard argument that a product of quotients is a quotient of the product and a subobject of a quotient is a quotient of a subobject so that $\underline{W}'$ is itself a variety. If $F \in \underline{W}'$ is free in $\underline{W}'$, it is a quotient of a $V \in \underline{V}$. Since F is free, the quotient map $V \longrightarrow F$ splits and hence F is a subobject of V and hence $F \in \underline{V}$. Thus $\underline{V}$ is a semi-variety in $\underline{W}'$. Actually $\underline{W}'$ is unique since every object in $\underline{W}'$ is a quotient of an object in $\underline{V}$. The same argument would show that the variety in which a semi-variety is embedded is also unique. <u>Henceforth we will understand $\underline{\underline{V}}$ to be a semi-variety, $\underline{\underline{W}}$ a containing variety and that $\underline{\underline{V}}$ contains the free algebras of $\underline{\underline{W}}$</u> .

(1.9) Now suppose that the theory Ω is commutative. That means that for any n-ary operation ω, m-ary operation ζ and algebra with underlying set S, the square

$$\begin{array}{ccc} S^{n\times m} \cong S^{m\times n} & \xrightarrow{\zeta^n} & S^n \\ \omega^n \downarrow & & \downarrow \omega \\ S^m & \xrightarrow{\zeta} & S \end{array}$$

commutes. The isomorphism in the upper left is most easily described by thinking of ω and ζ as acting on row vectors, $S^{n\times m}$ and $S^{m\times n}$ as $n\times m$ and $m\times n$ matrices respectively and the isomorphism as transpose. The implication of the commutability of this square is that ζ is a homomorphism with respect to ω (or vice versa).

If this holds for all ω,ζ the result is that each operation is not merely a function on the underlying sets but is in fact a morphism in $\underline{V}$.

(1.10) Suppose in addition that $\underline{V}_0$ is a variety for a commutative theory. If V' and V are algebras, (f_i), $i\epsilon n$ a family of morphisms $V' \longrightarrow V$ and ω an n-ary operation, the composite

$$V' \xrightarrow{(f_i)} V^n \xrightarrow{\omega} V$$

is again a morphism. This defines an operation $\omega(V',V)$ on $\mathrm{Hom}(V',V)$ and it is immediate that $\mathrm{Hom}(V',V)$ is again an algebra which we call $\underline{V}(V',V)$. It is easily seen that $\underline{V}(V,-)$ preserves limits and, by the adjoint functor theorem, has an adjoint $-\otimes V$. Finally let I be the free algebra on one generator. Then the required natural transformations and equivalences may easily be constructed to show that we have a symmetric monoidal closed category. This may be described as the canonical closed structure corresponding to a commutative theory.

(1.11) Even in case that $\underline{V}_0$ is only a semi-variety - but for a commutative theory - we can still make it into a closed category. In fact, let $\underline{W}_0$ be the category of algebras for the theory. Then $\underline{W}_0$ has the standard structure as above and we will see that $\underline{V}_0$ is an exponential ideal. In fact, let $V\epsilon\underline{V}_0$ and $W\epsilon\underline{W}_0$. Then W has a presentation as a coequalizer

$$FS \rightrightarrows FT \longrightarrow W$$

where FS and FT are the free algebras on the sets S and T respectively, and, as observed in (1.7), belong to $\underline{V}_0$. Then there is an equalizer

$$\begin{array}{ccccc}
\underline{W}(W,V) \longrightarrow & \underline{W}(FT,V) & \rightrightarrows & \underline{W}(FS,V) \\
 & \downarrow\cong & & \downarrow\cong \\
\underline{W}(W,V) \longrightarrow & V^T & \rightrightarrows & V^S
\end{array}$$

But since $V\epsilon\underline{V}_0$ and $\underline{V}_0$ is a quasi-variety, V^T and V^S and hence $\underline{W}(W,V)$ all belong to V.

(1.12) Whether or not the theory is commutative, it is possible that there is a symmetric monoidal closed structure on the semi-variety $\underline{V}$ which differs from the standard one. Here is the example which interests us. Let Π be a group, K a field and $\underline{V}$ be the category of representations of Π on K-vector spaces, otherwise known as $K[\Pi]$-modules. The theory of this variety is commutative according as Π is, but regardless there is a closed symmetric monoidal structure. For $\underline{V}(V',V)$ take the set of K-linear maps with Π action $(xf)(v') = xf(x^{-1}v')$ for $x\epsilon\Pi$, $f:V' \longrightarrow V$ and $v'\epsilon V'$. For $V'\otimes V$ we take the tensor product over K with $x(v\otimes v') = xv\otimes xv'$. The unit object I is just K with each group element acting by the identity map. This is not the free object on one generator which means that the functor $\mathrm{Hom}(I,-)$ is not the usual underlying set functor.

(1.13) Suppose that $\underline{V}$ is a semi-variety and also a closed symmetric monoidal category.

The varietal structure gives an underlying functor $\langle - \rangle$ which is represented by the free algebra on one generator which we denote J. We wish to describe a coherence between the internal and external hom which will say that up to natural isomorphism

$$\mathrm{Hom}(V',V) \subset \langle \underline{V}(V',V) \rangle \subset \mathrm{Hom}(\langle V' \rangle , \langle V \rangle) .$$

and that these inclusions cohere with the unitary and associative maps for functions between sets. The existence of the requisite functions can be shown to follow from the hypothesis that J be a cocommutative coalgebra object in $\underline{V}$. The first map is an injection provided the counit map is an epimorphism. What additional hypothesis is needed to force the second to be an injection is not clear. Nor does the question seem worth pursuing at this stage in the theory. The main point to keep in mind is that these objects come provided with underlying sets that are very much part of their structure and that both the external hom and the elements of the internal hom are functions between those underlying sets and compare the way functions do. For later use, we call an element of $\langle \underline{V}(V',V) \rangle$ a <u>pseudomap</u> from $V' \longrightarrow V$.

(1.14) The fact that we are supposing that $\mathrm{Hom}(V,V')$ and $\langle \underline{V}(V,V') \rangle$ are contained in $\mathrm{Hom}(\langle V\rangle, \langle V'\rangle)$ - and for future reference $\underline{V}(A,B) \subset \mathrm{Hom}(\langle |A| \rangle, \langle |B| \rangle)$ - makes for vast simplifications in our theory. Specifically, we will in the future, say things like,"There is a canonical map $\underline{V}(A,(B,E)) \longrightarrow \underline{V}(B,(A,E))$." What this means will of course depend on the exact statement. But in all cases it means that at the underlying set level it comes down to a well-known canonical map, in this case the transposition between a map $\langle |A| \rangle \longrightarrow \mathrm{Hom}(\langle |B| \rangle, \langle |E| \rangle)$ and a map $\langle |B| \rangle \longrightarrow \mathrm{Hom}(\langle |A| \rangle, \langle |E| \rangle)$. In these notes, this will be understood without further mention. The alternative is to make the statements - not to mention the proofs - of most of the propositions unbearably awkward. An added advantage is that all coherence questions vanish. This is not the reason for adopting a rigid grounding functor (that is basically to be able to apply (2.5) below and to make sense of the idea of a convergence uniformity) but as long as we seem to be stuck with one, we may as well take advantage of it.

(1.15) <u>Example</u>. Here is the example which we will continue throughout the chapters devoted to the theory. Let K be a (commutative) field and $\underline{V}$ denote the category of vector spaces over K. Then $\underline{V}$ is already a variety. It has the nullary 0, unary and binary operations and equations required to define an abelian group as well as a unary operation, multiplication by λ, for each $\lambda \epsilon K$. It is well known to be a closed category. In fact, the theory is commutative. If V and V' are vector spaces and $f,g : V \longrightarrow V'$ are maps, $f+f' : V \longrightarrow V'$ is defined by $(f+f')(v) = f(v)+f'(v)$. Similarly, for $\lambda \epsilon K$, $(\lambda f)(v) = \lambda f(v)$ defines the operation of λ on $\mathrm{Hom}(V,V')$. The unit object for this hom is K and the tensor product is the usual one.

2. <u>Uniform Spaces.</u>

(2.1) Let S be a set. Let $\underline{u} = (S_\beta)$ be a cover of S, that is a collection of subsets of S whose union is S. For $x \epsilon S$, let $\underline{u}^*(x) = \cup\{u \epsilon \underline{u} \mid x \epsilon u\}$. Let $\underline{u}^* = \{\underline{u}^*(x) \mid x \epsilon S\}$.

If $\underline{u}$ and $\underline{v}$ are covers, we say that $\underline{u}$ refines $\underline{v}$ if for $x \in v \in \underline{v}$ there is a $u \in \underline{u}$ such that $x \in u \subset v$. Then a <u>uniform structure</u> on S is a collection $\underline{U}$ of covers such that $\underline{u} \in \underline{U}$ implies the existence of a $\underline{v} \in \underline{U}$ such that $\underline{v}^*$ refines $\underline{u}$. Such a $\underline{v}$ is called a <u>star refinement</u> of $\underline{U}$. A pair $(S,\underline{U})$ consisting of a set S and a uniform structure $\underline{U}$ on S will be called a <u>uniform space</u>. The covers in $\underline{U}$ are called <u>uniform covers</u>.
If $(S_1,\underline{U}_1)$ and $(S_2,\underline{U}_2)$ are uniform spaces a map $f:S_1 \longrightarrow S_2$ is called <u>uniform</u> if for all $\underline{u}_2 \in \underline{U}_2$ there is $\underline{u}_1 \in \underline{U}_1$ which refines $f^{-1}(\underline{u}_2)$. A uniform structure $\underline{U}$ on S is called <u>separated</u> if for $x \neq y$ in S there is a $\underline{u} \in \underline{U}$ such that no set in $\underline{u}$ contains both x and y. <u>From now on we will suppose all uniform spaces are separated</u>. For a thorough discussion of uniform spaces including the relation between the definition given here and that given by entourages, see [Isbell] , especially problem 8 of chapter I and pp. 28-29 of $X = (S,\underline{U})$ is a uniform space, write $S = |X|$ for the underlying set.

(2.2) If $(S,\underline{U})$ is a uniform space, there is associated a topological space (S,τ) with the same point set. Take as a neighbourhood base at x all the sets $\underline{u}^*(x)$ where $\underline{u} \in \underline{U}$. This topology is called the uniform topology. Every uniform map induces a continuous one on the associated topological spaces but not conversely. In particular two distinct uniform structures may induce the same topology. For instance, as topological spaces, $\mathbb{R} \cong (-1,1)$. As uniform spaces they are inequivalent since the former is complete (see below) but the latter is not.

(2.3) If $(S,\underline{U})$ is a uniform space and S' is a subset of S, there is a natural uniformity induced on S'. Namely the collection of all $u \cap S'$ for $u \in \underline{u} \in \underline{U}$. The uniform topology for the induced uniformity on S' is the same as that induced by the uniform topology on S.

We say that the uniform space $(S_1, \underline{U}_1)$ is complete if for every uniform space $(S,\underline{U})$, every topologically dense subset $S' \subset S$ and every function $f':S' \longrightarrow S_1$ which is uniform in the induced uniformity, there is a uniform $f:S \longrightarrow S_1$ whose restriction to S' is f'.

(2.4) It is shown in [Isbell] , chapter II how every uniform space can be embedded (i.e. has the induced uniformity) as a dense subset of a complete space and that this embedding is unique up to a unique isomorphism. We denote this completion of $(S,\underline{U})$ by $(S,\underline{U})^{\sim}$ or just by $S^{\sim}$ if $\underline{U}$ is understood.

(2.5) <u>Lemma</u>. Suppose

$$\begin{array}{ccc} (S_1,\underline{U}_1) & \xrightarrow{f_1} & (S_2,\underline{U}_2) \\ g \downarrow & & \downarrow h \\ (T_1,\underline{V}_1) & \xrightarrow{f_2} & (T_2,\underline{V}_2) \end{array}$$

is a commutative diagram of uniform spaces such that f_1 is the inclusion of a dense subset, equipped with the induced uniformity, f_2 is an isomorphism of underlying sets and $\underline{V}_1$ has a basis consisting of sets whose image under f_2 is closed in the uniform topology of $\underline{V}_2$. Then there is a unique map $(S_2,\underline{U}_2) \longrightarrow (T_1,\underline{V}_1)$ making both triangles commute.

Proof. Although I have tried to make these notes largely self-contained, this argument requires more than a passing acquaintance with the theory of uniform spaces to be understood. For this I refer to [Isbell]. Otherwise, it can best be appreciated by imagining the spaces to be metric and replacing Cauchy nets by Cauchy sequences. n

We may suppose without loss of generality that $T_1 = T_2 = T$ and f_2 is the identity map which means that $\underline{V}_1$ is a refinement of $\underline{V}_2$. Then given that h is uniform with respect to $\underline{V}_2$ we must show it is with respect to $\underline{V}_1$. Suppose now that $s \in S_2$ and $\{s_\alpha\}$ is a Cauchy net in S_1 which converges to s. Then $\{hs_\alpha\}$ converges to hs. Since g is uniform, $\{gs_\alpha\}$ is a Cauchy net with respect to $\underline{V}_1$. Now let $\underline{v}_1 \in \underline{V}_1$. There is $\underline{v}_1' \in \underline{V}_1$ such that $\underline{v}_1'^*$ refines $\underline{v}_1$ and by the hypothesis we may also suppose that the sets in v_1' are closed in the uniform topology determined by $\underline{V}_2$. There is a set $V \in \underline{v}_1'$ and an index β such that $\gamma \geq \beta$ implies $gs_\gamma \in V$ (that is just from the definition of Cauchy net). Now $\{gs_\gamma\}, \gamma \geq \beta$ is again a Cauchy net in $\underline{V}_2$ and converges, in the uniform topology of $\underline{V}_2$ to hs. Since V is closed in that topology and every $gs_\gamma \in V$, it follows that $hs \in V$ as well. Thus $V \subset \underline{v}_1'^*(hs)$. Hence for all $\gamma > \beta, gs_\alpha \in v_1'^*(hs)$. Thus in every $\underline{V}_1$ cover of T the net is eventually in one of the neighbourhoods of hs and thus converges to hs.

This does not quite show that h is uniform to $\underline{V}_1$ but we may argue as follows. The map g has a uniform extension to $g^\sim : (S_2,U_2) \longrightarrow (T,V_1)^\sim$ constructed by choosing for each $s \in S_2$ a Cauchy net $\{s_\alpha\}$ of elements of S_1 which converges to it. Then as above $\{gs_\alpha\}$ is Cauchy in $(T,\underline{V}_1)$ and converges to a unique point, $g^\sim(s)$, of $(T,\underline{V}_1)^\sim$. But we have just shown that all such points already lie in T.

(2.6) A metric space (M,α) is a uniform space in a natural way. In fact the covers by ε-spheres, $\varepsilon>0$ determine a uniformity called the metric uniformity. It follows from [Isbell], I.14 that for any uniform space $X = (S,\underline{U})$ and any $\underline{u} \in \underline{U}$ there is a metric space (M,α) and a uniform map $f: X \longrightarrow (M,\alpha)$ such that the inverse image of the 1-spheres refines $\underline{u}$. Such an f is called a pseudometric for X. If $\underline{u}$ ranges over a basis of $\underline{U}$ a corresponding collection of f is called a basis of pseudometrics for X.

(2.7) Suppose $(S,\underline{U})$ and $(T,\underline{V})$ are uniform spaces and F is a set of functions $S \longrightarrow T$. We say that F is an equiuniform set if for all $\underline{v} \in \underline{V}$ there is a single $\underline{u} \in \underline{U}$ which refines $f^{-1}(\underline{v})$ for all $f \in F$. It is easy to see that every finite set of uniform functions is an equiuniform set.

(2.8) If, on the other hand, S is a set and $(T,\underline{V})$ is a uniform space, a collection Φ of sets of functions $S \longrightarrow T$ determines a uniformity $\underline{U}$ on S such that the sets in Φ are equiuniform. This is most readily described in terms of a basis $\underline{d}$ of pseudometrics for $\underline{V}$. Let

$$\Gamma(s,F,d) = \{s' \mid d(fs,fs') < 1 \quad \text{for all} \quad f\in F\}$$

for $s\in S, F\in\Phi, d\in \underline{d}$. Then the sets

$$\{\Gamma(s,F,d) \mid s\in S\}$$

are a cover of S which determine a uniform structure as F and d vary. A basis of pseudometrics for this structure is given by the functions $d\cdot F$ where $(d\cdot F)(s,s') = \sup\{d(fs,fs') \mid f\in F\}$.

(2.9) We can instead suppose given a collection Σ of subsets of S and a set F of functions $S \longrightarrow T$. In that case we can think of Σ as a collection of sets of functions $F \longrightarrow T$. Then the discussion above describes a uniform structure on F . We will call a uniformity so described a convergence uniformity. A special case is $\Sigma = \{S\}$ in which case F has the uniformity of uniform convergence on all of S that is considered in [Isbell], Chapter III .

(2.10) If $X \longrightarrow Y$ is an injective uniform map we say it is embedding provided that X is uniformly isomorphic to its image when that image is given the induced uniformity. We note the obvious fact that the category of uniform spaces has a factorization system consisting of surjections and embeddings as the epimorphic and monomorphic parts, respectively. In particular, the embeddings possess the usual properties of invariance under composition and left cancellation (see [Kelly]).

(2.11) If $\{X_\omega\}$ is a collection of uniform spaces, every pseudometric on $X = \Pi X_\omega$ is majorized by a pseudometric constructed as follows. Let $i=1, \ldots , n$ be a finite number of indices and $p_i : X \longrightarrow X_i$ be the corresponding projective map. Let d_i be a pseudometric in X_i and let

$$d(x,y) = \sup(d_1(p_1x,p_1y), d_2(p_2x,p_2y), \ldots, d_n(p_nx,p_ny)) .$$

In the sequel we will write $d = \sup\{d_i(p_i,p_i)\}$. It is easy to show that the uniformity defined by the pseudometrics so defined is the coarsest for which the projections are uniform.

(2.12) Proposition. Let $\{X_\omega\}$ be a collection of uniform spaces and X have the discrete uniformity. Then any map $\Pi X_\omega \longrightarrow X$ factors through a finite product.

Proof. X has a uniform cover by singletons. The inverse image of that cover is required to be a uniform cover of the product. A uniform cover of a product is refined by a cover of the following form. Let $1,\ldots,n$ be a finite number of indices, $\underline{u}_1,\ldots,\underline{u}_n$ be covers in the corresponding space. Then the collection of all sets ΠV_ω where $V_\omega \in \underline{u}_\omega$, $\omega=1,\ldots,n$ and $V_\omega = X_\omega$ otherwise determines a cover. That f takes every such set into a single point implies, in particular, that if $x_\omega = x'_\omega$ except for $\omega=1,\ldots,n$, then $f((x_\omega)) = f((x'_\omega))$, or, in other words, that f factors through $\Pi X_\omega \longrightarrow X_1 \times \ldots \times X_n$.

(2.13) Notes on uniform spaces. There are really three definitions of uniform spaces. The first is by the uniform covers, given here. The second is by families of seminorms and the third by entourages (see [Kelley], Chapter VI). The first is the most intuitively geometric as well as the most useful for things like topological groups where each neighborhood of 1 gives rise to the cover gotten by translating that neighborhood all over the group. This automatically defines a uniformity - the continuity of multiplication is equivalent to the existence of star refinements - called the canonical uniformity. A group homomorphism is continuous iff it is uniform. Thus there is no distinction in that case between topological and uniform notions. The definition by pseudonorms seems the most useful in these notes. The definition by entourages is probably most useful in connection with compactness arguments. A compact Hausdorff space has a canonical - and unique - uniformity for which the entourages are all neighborhoods of the diagonal. Every map to another uniform space is continuous iff it is uniform. At any rate there is no "best" definition. It is important that all three are equivalent.

3. Uniform Space Objects

(3.1) Let $\underline{V}$ be a semivariety with underlying set functor $\langle - \rangle$. By a preuniform structure on an object $V \in \underline{V}$ we mean a uniform structure on $\langle V \rangle$ such that all the operations are uniform when the powers of $\langle V \rangle$ are equipped with the product uniformity. (see [Isbell] , Chapter II.)

If V is a preuniform object in $\underline{V}$ let $\langle V \rangle$ denote the underlying uniform space and $|V|$ denote the underlying discrete object of $\underline{V}$. Both $|\langle V \rangle|$ and $\langle |V| \rangle$ are the same underlying set of V .

(3.2) If V has a preuniform structure $\underline{U}$, we say that V is completable (in $\underline{V}$) provided the uniform completion of V also lies in $\underline{V}$. To explain this more clearly, consider the case that $\underline{V}$ is a variety. Then V is dense in $V^{\sim}$ (we omit the < >) and it is readily seen that for any n , V^n is dense in $(V^{\sim})^n$. For any n-ary operation σ , the map $V^n \xrightarrow{\sigma} V \to V^{\sim}$ has, then, an extension to a map $V^{\sim n} \to V^{\sim}$. Any equation required to be satisfied by an algebra of $\underline{V}$ is satisfied when restricted to the dense subset V . But an equation is satisfied exactly when two maps are equal and two uniform maps which agree on a dense subset are equal. In the general case, what is at issue is that $\tilde{V}$ belong to the semi-variety $\underline{V}$.

(3.3) Now suppose that $\underline{V}$ is closed (by which we understnad that the situation is that exposed in section 1, in particular that we have a symmetric monoidal structure). We say that V is admissible provided for all preuniform objects V', there is a subobject $\mathrm{Un}\underline{V}(V',V)$ of $\underline{V}(V',V)$ such that

$$\begin{array}{ccc} \langle \mathrm{Un}\underline{V}(V',V) \rangle & \longrightarrow & \mathrm{Un}(\langle V' \rangle, \langle V \rangle) \\ \downarrow & & \downarrow \\ \langle \underline{V}(|V'|, |V|) \rangle & \longrightarrow & \mathrm{Hom}(\langle |V'| \rangle, |\langle V \rangle|) \end{array}$$

is a pullback. Here, of course, $\mathrm{Un}(\langle V' \rangle, \langle V \rangle)$ is the set of uniform maps between these sets.

(3.4) If the theory of the semi-variety $\underline{V}$ is the commutative theory $\underline{Th}$ and $\underline{V}$ has the induced closed monoidal structure, it is easy to see that $\mathrm{Un}(\langle V' \rangle \langle V \rangle) \cap \mathrm{Hom}(V',V)$ is a $\underline{Th}$-algebra. For $\omega V = \omega : V^n \to V$ is an operation which we have supposed to be uniform, then for any collection $f_i : V' \longrightarrow V$, $i \in n$, of uniform morphisms, the composite

$$V' \xrightarrow{(f_i)} V^n \xrightarrow{\omega} V$$

is also a uniform morphism. But this is $\omega(f_i)$, the Th operation in $\underline{V}(V',V)$. Thus the admissibility of V comes down to whether or not that particular Th-algebra belongs to $\underline{V}$.

(3.5) Even when Th is not necessarily commutative and $\underline{V}$ has some *ad hoc* closed monoidal structure, there is at most one candidate for $\mathrm{Un}\,\underline{V}(V',V)$. For suppose V_0 and V_1 are two subobjects of an object V_2 such that $< V_0 > = < V_1 >$ as subobjects of $< V_2 >$. Then both inclusions $V_0 \cap V_1 \longrightarrow V_0$ and $V_0 \cap V_1 \longrightarrow V_1$ become equalities at the underlying set. Since the underlying set functor reflects isomorphisms it follows that $V_0 = V_0 \cap V_1 = V_1$. Thus the subobject $\mathrm{Un}\,\underline{V}(V',V)$ is unique, provided it exists.

(3.6) We say that V is a uniform object if it is a preuniform object, if it is admissible, completable and its completion is adimissible. We let $\mathrm{Un}\,\underline{V}$ denote the category of uniform $\underline{V}$ objects, with $< \mathrm{Un}\,\underline{V}(-,-) >$ as the hom. From the definition it follows that a morphism is simply one which is both a morphism in $\underline{V}$ and uniform on the underlying sets.

(3.7) If $\underline{V}$ is the category of all the algebras for a commutative theory with the natural closed monoidal structure thereby induced, then as observed in (3.4) every preuniform structure is admissible and as observed in (3.2) every preuniform structure is completable. Thus there is, in that case no distinction between preuniform and uniform. Even in that case, it is not true, unless the theory is finitary, that every object in $\underline{V}$ becomes a uniform object when equipped with the discrete uniformity. In fact, it follows from 2.12 that a map from a product of discrete spares to a discrete space is uniform iff it is a function of only finitely many coordinates.

(3.8) Let $\{U_\omega\}$ be a collection of uniform objects. Let $U = \Pi U_\omega$ equipped with the product uniformity - the coarsest for which the projections are uniform. I claim that U is a uniform object. To see this, let V be a preuniform object. Then since

$$\begin{array}{ccc} <\mathrm{Un}\underline{V}(V,U_\omega)> & \longrightarrow & \mathrm{Un}(<V>,<U_\omega>) \\ \downarrow & & \downarrow \\ <\underline{V}(|V|,|U_\omega|)> & \longrightarrow & \mathrm{Hom}(<|V|>,<|U_\omega|>) \end{array}$$

is a pullback, for each i and both $|-|$, $<->$ and the hom functors commute with products, so is

$$\begin{array}{ccc} <\Pi\underline{V}(V,U_\omega)> & \longrightarrow & \mathrm{Un}(<V>,<\Pi U_\omega>) \\ \downarrow & & \downarrow \\ <\underline{V}(|V|,|\Pi U_\omega|)> & \longrightarrow & \mathrm{Hom}(<|V|>,<|\Pi U_\omega|>) \end{array}$$

which shows that U is admissible. Moreover U_ω is dense in $U_\omega^\sim$ from which U is dense in $\Pi U_\omega^\sim$ and the latter is complete so that $U^\sim = \Pi U_\omega^\sim$ and U is completable. Evidently $U^\sim$ is admissible being a product of admissible objects. It is evident that U is the product of the U_ω .

(3.9) Now let $U' \rightrightarrows U''$ be two maps of uniform objects and U be their equalizer, given the structure as a subspace of U'. Exactly the same argument as above suffices to show that U is admissible. To show that it is completable is, however, another matter entirely. It is easy to see that $U^\sim$ is a subobject of $U'^\sim$ but there does not seem to be any obvious reason that it is a regular subobject. If we supposed that $\underline{V}$ was a quasi-variety - closed under products and subobjects - that would settle it. But in the one example we have in which $\underline{V}$ is not a variety - that of Banach spaces (see IV. 3.) - $\underline{V}$ is not a quasi-variety either. What happens there is that any closed subobject of a uniform object is a uniform object. Thus the appropriate hypothesis at this point is unclear for want of examples and we leave it as an open question.

(3.10) Proposition. The underlying functor $|\ | : \mathrm{Un}\underline{V} \longrightarrow \underline{V}$ creates products. Provided the completion of an equalizer is an admissible pre-uniform object of $\underline{V}$, the functor creates limits. If $\underline{W}$ is the category of all algebras, this condition is satisfied provided $\underline{V}$ is closed in $\underline{W}$ under all subobjects or at least if the $\underline{W}$ object underlying a closed $\mathrm{Un}\underline{W}$ subobject of an object in $\mathrm{Un}\underline{V}$ belongs to $\underline{V}$.

This last condition means that if $U \in \mathrm{Un}\underline{V}$, and $\langle U' \rangle \subset \langle U \rangle$ is a closed subspace of the uniform space with $|U'| \in \underline{W}$, then $|U'| \in \underline{V}$.

In the sequel, we will simply suppose that the conclusion of this proposition is automatically satisfied.

(3.11) Let $V \in \underline{V}$ and $U \in \mathrm{Un}\underline{V}$. We let $[V, U]$ denote $\underline{V}(V, |U|)$ equipped with the coarsest uniformity for which the map evaluation at $v: [V,U] \longrightarrow U$ is uniform for each $v \in V$. Equivalently we require that

$$\langle [V,U] \rangle \longrightarrow \langle U \rangle^{\langle V \rangle}$$

be a uniform embedding. Then $[V,U]$ is certainly a preuniform object. Now I claim that

$$\begin{array}{ccc}
\langle \underline{V}(V, \mathrm{Un}\underline{V}(U',U)) \rangle & \longrightarrow & \underline{\mathrm{Un}}(\langle U' \rangle, \langle [V,U] \rangle) \\
\downarrow & & \downarrow \\
\langle \underline{V}(|U'|, |[V,U]|) \rangle & \longrightarrow & \mathrm{Hom}(\langle |U'| \rangle, \langle |[V,U]| \rangle)
\end{array}$$

is a pullback. We begin with the fact that

$$\begin{array}{ccc}
\langle \mathrm{Un}\underline{V}(U',U) \rangle & \longrightarrow & \mathrm{Un}(\langle U' \rangle, \langle U \rangle) \\
\downarrow & & \downarrow \\
\langle \underline{V}(|U'|, |U|) \rangle & \longrightarrow & \mathrm{Hom}(\langle |U'| \rangle, \langle |U| \rangle)
\end{array}$$

is a pullback in $\underline{S}$. Applying $\mathrm{Hom}(\langle V \rangle, -)$ we see that

$$\begin{array}{ccc}
\mathrm{Hom}(\langle V \rangle, \langle \mathrm{Un}\underline{V}(U',U) \rangle) & \longrightarrow & \mathrm{Hom}(\langle V \rangle, \mathrm{Un}(\langle U' \rangle, \langle U \rangle)) \\
\downarrow & & \downarrow \\
\mathrm{Hom}(\langle V \rangle, \langle \underline{V}(|U'|, |U|) \rangle) & \longrightarrow & \mathrm{Hom}(\langle V \rangle, \mathrm{Hom}(\langle |U'| \rangle, \langle |U| \rangle))
\end{array}$$

is as well. Now since $\langle V \rangle$ is discrete

$$\mathrm{Hom}(\langle V \rangle, \mathrm{Un}(\langle U' \rangle, \langle U \rangle)) \cong \mathrm{Un}(\langle U' \rangle, \langle U \rangle^{\langle V \rangle})$$

and

$$\mathrm{Hom}(\langle V\rangle, \mathrm{Hom}(\langle|U'|\rangle, \langle|U|\rangle)) \cong \mathrm{Hom}(\langle|U'|\rangle, \langle|U^{\langle V\rangle}|\rangle).$$

since by

$$\begin{array}{ccc} \langle \underline{V}(V,\mathrm{Un}\underline{V}(U',U))\rangle & \longrightarrow & \mathrm{Hom}(\langle V\rangle, \langle \mathrm{Un}\,\underline{V}(U',U)\rangle) \\ \downarrow & & \downarrow \\ \langle \underline{V}(V,\underline{V}(|U'|,|U|)\rangle & \longrightarrow & \mathrm{Hom}(\langle V\rangle, \underline{V}(|U'|,|U|)) \end{array}$$

is a pullback, it follows that so is

$$\begin{array}{ccc} \langle \underline{V}(V,\mathrm{Un}\underline{V}(U'U))\rangle & \longrightarrow & \underline{\mathrm{Un}}(\langle U'\rangle\langle U^{\langle V\rangle}\rangle) \\ \downarrow & & \downarrow \\ \langle \underline{V}(V,\underline{V}(|U'|,|U|)\rangle & \longrightarrow & \mathrm{Hom}(\langle|U'|\rangle, \langle|U^{\langle V\rangle}|\rangle) . \end{array}$$

If $f:V \longrightarrow \mathrm{Un}\,\underline{V}(U',U) \subset \underline{V}(U',U)$ is a pseudomap, then $f\in \langle \underline{V}(V,\underline{V}(U,U))\rangle \cong \langle \underline{V}(U',\underline{V}(V,U))\rangle$ so f is a pseudomap $U' \longrightarrow \underline{V}(V,U) = |[V,U]|$. Since it also determines a uniform map $\langle U'\rangle \longrightarrow \langle U^{\langle V\rangle}\rangle \cong \langle U\rangle^{\langle V\rangle}$ it gives a uniform pseudomap $U' \longrightarrow [V, U]$. That is, the image of the upper map in the last square lies in $\mathrm{Un}(\langle U'\rangle, \langle [V,U]\rangle)$. In a similar way the image of the lower map lies in $\mathrm{Hom}(\langle|U'|\rangle, \langle|[V,U]|\rangle)$ which gives the desired result.

(3.12) This not only shows that $[V,U]$ is admissible but gives the desired adjunction for a cotensor $\mathrm{Un}\,\underline{V}(U',[V,U]) \cong \underline{V}(V,\mathrm{Un}\,\underline{V}(U',U))$. However, we still have to show that $[V,U]$ is completable and that its completion is admissible. First we consider the case that U is complete. Then for U' dense in U'', we know (essentially because we have assumed it) that

$$\mathrm{Un}\,\underline{V}(U'',U) \longrightarrow \mathrm{Un}\,\underline{V}(U',U)$$

is an isomorphism. Applying $\underline{V}(V,-)$ we see that

$$\underline{V}(V,\mathrm{Un}\underline{V}(U'',U)) \longrightarrow \underline{V}(V,\mathrm{Un}\underline{V}(U',U))$$

or

$$\mathrm{Un}\underline{V}(U'',[V,U] \longrightarrow \mathrm{Un}\underline{V}(U',[V,U])$$

is an isomorphism. This implies that $[V,U]$ is complete as well. Now if U is arbitrary we use the hypothesis of (3.10) to infer that the closure of $[V,U]$ in $[V,\tilde{U}]$ is an admissible preuniform $\underline{V}$-object as well. To see that we need only observe that $[V,U]$ is uniformized as a subobject of $[V,\tilde{U}]$. More generally, we have the following

(3.13) Proposition. Let U be embedded in U'. Then $[V,U]$ is embedded in $[V,U']$.
Proof. In the square

$$\begin{array}{ccc} \langle [V,U]\rangle & \longrightarrow & \langle U\rangle^{\langle V\rangle} \\ \downarrow & & \downarrow \\ \langle [V,U]\rangle & \longrightarrow & \langle U'\rangle^{\langle V\rangle} \end{array}$$

the upper and right hand map are embeddings. Thus the composite and therefore the left hand map is one as well.

(3.14) One more hypothesis will have to be made. When $\underline{V}(I,-)$ represents the underlying set functor, there is no difference between maps and pseudomaps. The fact that the operations in the theory are uniform implies that the uniform limit of maps is a map. More specifically, if $A_1 \longrightarrow A_2$ is a dense embedding and B is complete, any map $A_1 \longrightarrow B$ extends to a map $A_2 \longrightarrow B$. Now we must suppose the same to hold for pseudomaps. It follows immediately that there is a canonical map $\underline{A}(A,B) \longrightarrow \underline{A}(\tilde{A},\tilde{B})$.

(3.15) <u>Example</u>. If $\underline{V}$ is the category of vector spaces, it is a variety with a commutative theory. Thus the category $Un\underline{V}$ is simply the category of preuniform objects in $\underline{V}$. Moreover, according to the remarks on groups in (2.12) a uniform object is the same as a topological vecotr space over the discrete field K. This is the same as a topological group which is simultaneously a vector space such that all scalar multiplications are continuous.

4. *-Autonomous Categories

(4.1) From here on, we revive an old name due to Linton and call a symmetric monoidal closed category <u>autonomous</u>. By a *-autonomous category is meant

(i) An autonomous category $\underline{G}$;

(ii) A closed functor $(-)^* : \underline{G}^{op} \longrightarrow \underline{G}$;

(iii) An equivalence $d = dG : G \longrightarrow G^{**}$.

This is subject to one axiom, that the diagram

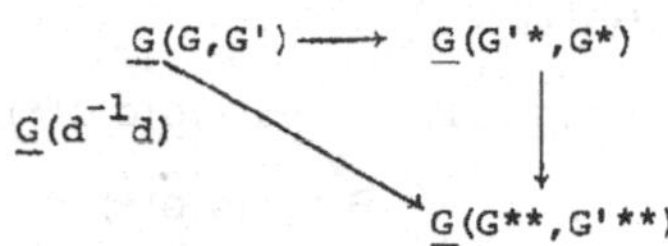

whose horizontal and vertical arrow are the actions of $(-)^*$ on the internal hom, commute.

(4.2) An immediate result is: the horizontal arrow is a split mono and the vertical one a split epi. But the latter is an instance of the former and thus is also a split mono which implies that both are isomorphisms.

(4.3) In fact, far less data than that is required to have a *-autonomous category. Suppose we have a monoidal category $\underline{G}$ with unit I and tensor product $\otimes$ equipped with a functor $\underline{G}^{op} \longrightarrow \underline{G}$ which is full and faithful and such that there is natural equivalence $Hom(G_1 \otimes G_2, G_3^*) \longrightarrow Hom(G_1,(G_2 \otimes G_3)^*)$. In that case, define $\underline{G}(G',G) = (G' \otimes G^*)^*$. Let $T = I^*$. Then we see that there is a 1-1 consequence between maps $G \otimes G' \longrightarrow T$ and $G' \longrightarrow (G \otimes I)^* \cong G^*$. Since the tensor is symmetric, there is similarly a correspondence between maps $G' \otimes G \longrightarrow T$ and $G' \longrightarrow G^*$. Putting these together gives a correspondence between $G' \longrightarrow G^*$ and $G \longrightarrow G'^*$. Since $(-)^*$ is full and faithful we have correspondences

$$G \longrightarrow G'$$
$$G'^* \longrightarrow G^*$$
$$G \longrightarrow G'^{**}$$

and by the Yoneda lemma $G' \cong G'^{**}$.

Next define $\underline{G}(G',G) = (G'\otimes G^*)^*$. There is a 1-1 correspondence between maps

$$G_1 \longrightarrow \underline{G}(G_2,G_3) = (G_2\otimes G_3^*)^*$$
$$G_2\otimes G_3^* \longrightarrow G_1^*$$
$$G_3^*\otimes G_2 \longrightarrow G_1^*$$
$$G_3^* \longrightarrow (G_2\otimes G_1)^*$$
$$G_2\otimes G_1 \longrightarrow G_3$$
$$G_1\otimes G_2 \longrightarrow G_3$$

which demonstrates that $\underline{G}(G_2,-)$ is left adjoint to $-\otimes G_2$. It follows from [Eilenberg-Kelly] , II.3 especially the material following proposition 3.1, that $\underline{G}$ then becomes autonomous with this definition of hom. The equivalence of $\underline{G}(G',G)$ and $\underline{G}(G^*,G'^*)$ is immediate. Note that this is independent of whether or not $(-)^*$ is a monoidal functor. That seems to be more or less equivalent to $I^* \cong I$, which holds in all the examples but one. Thus the data at the beginning of this section suffice to determine a *-autonomous category.

(4.4) On the other hand suppose that the category $\underline{G}$, internal hom functor $\underline{G}(-,-)$, unit object I together with required natural transformations and equivalences constitute a symmetric closed category. Suppose also that there is a full faithful functor $(-)^*$: $\underline{G}^{op} \longrightarrow \underline{G}$ and an equivalence $\mathrm{Hom}(G_1,\underline{G}(G_2,G_3^*)) \cong \mathrm{Hom}(G_3,\underline{G}(G_2,G_1^*))$. As before, let $T = I^*$. Then there is an equivalence

$$\mathrm{Hom}(G',\underline{G}(G,T)) \cong \mathrm{Hom}(I,\underline{G}(G,G'^*))$$
$$\cong \mathrm{Hom}(G,G'^*) \ .$$

Similarly,

$$\mathrm{Hom}(G',G^*) \cong \mathrm{Hom}(G',\underline{G}(I,G^*))$$
$$\cong \mathrm{Hom}(G,\underline{G}(I,G'^*))$$
$$\cong \mathrm{Hom}(G,G'^*) \ .$$

Comparing these, we see that $G^* = \underline{G}(G,T)$. Also, $\mathrm{Hom}(G',G^*) \cong \mathrm{Hom}(G,G'^*)$ which together with $\mathrm{Hom}(G',G) \cong \mathrm{Hom}(G^*,G'^*)$ implies that $(-)^*$ is an equivalence. In fact, we have $\mathrm{Hom}(G',G) \cong \mathrm{Hom}(G^*,G'^*) \cong \mathrm{Hom}(G',G^{**})$ while evidently $\mathrm{Hom}(G^*,G'^*) = \mathrm{Hom}(G'^{**},G^{**})$ and by the Yoneda lemma $G' \cong G'^{**}$.

(4.5) Since $G^* \cong \underline{G}(G,T)$ the internal composite gives a natural transformation

$$(G',G) \longrightarrow (G^*,G'^*)$$

which followed by the isomorphism above gives a map $(G',G) \longrightarrow (G',G^{**})$. It follows from coherence that the diagram

$$\begin{array}{ccc} \underline{G}(G',G) & \longrightarrow & \underline{G}(G^*,G'^*) \\ & \searrow & \downarrow \\ & & \underline{G}(G',G^{**}) \end{array}$$

commutes. Since the vertical and diagonal maps are isomorphisms, so is the horizontal one which means that $(-)^*$ is internally full and faithful. Now we define $G' \otimes G = \underline{G}(G', G^*)^*$. We see that there are natural 1-1 correspondences between maps

$$G_1 \otimes G_2 \longrightarrow G_3$$

$$\underline{G}(G_1, G_2^*)^* \longrightarrow G_3$$

$$G_3^* \longrightarrow \underline{G}(G_1, G_2^*)$$

$$G_1 \longrightarrow \underline{G}(G_3^*, G_2^*) \cong \underline{G}(G_2, G_3) \ .$$

Here we have freely used the equivalence between G and G^{**}. That we have a *-autononomous category now follows from [Eilenberg-Kelly], I.3 .

(4.6) The main purpose in these notes is to begin with a good deal less than a *-autononomous category and construct one. This leads us to define the notion of a <u>pre-*-autonomous situation</u>. This consists of a category $\underline{A}$, two full subcategories $\underline{C}$ and $\underline{D}$, an equivalence **of categories**

$$(-)^* : \underline{C}^{op} \longrightarrow \underline{D} \ ,$$

a functor $(-,-) : \underline{C}^{op} \times \underline{D} \longrightarrow \underline{D}$ and an object $I \in \underline{C}$. These are subject to the axioms of a *-autonomous category insofar as they make sense. In particular, we suppose that

(i) $(I, D) \cong D$;

(ii) $\mathrm{Hom}(I, (C, D)) \cong \mathrm{Hom}(C, D)$;

(iii) $(C_1, (C_2, C_3^*)) \cong (C_3, (C_2, C_1^*))$;

(iv) $\mathrm{Hom}(C^*, C') \cong \mathrm{Hom}(C'^*, C)$.

These are subject to certain coherence conditions which will be introduced as needed.

(4.7) Since $(-)^*$ is an equivalence, it has, up to a natural isomorphism, an inverse functor which we temporarily denote $(-)^{\#} : \underline{D}^{op} \longrightarrow \underline{C}$. Then (iii) above may be rewritten $(C, D) \cong (D^{\#}, C^*)$ which, by (ii) implies that $\mathrm{Hom}(C, D) \cong \mathrm{Hom}(D^{\#}, C^*)$. Similarly (iv) may be rewritten $\mathrm{Hom}(D, C) \cong \mathrm{Hom}(C^*, D^{\#})$. Should it happen that D also belongs to $\underline{C}$ we have, since $(-)^*$ is an equivalence, that

$$\mathrm{Hom}(D, D) \cong \mathrm{Hom}(D^*, D^{\#})$$

and

$$\mathrm{Hom}(D, D) \cong \mathrm{Hom}(D^{\#}, D^*) \ .$$

The coherences alluded to above would require that if $f \longmapsto \alpha(f)$ describes the map

$$\mathrm{Hom}(C, D) \xrightarrow{\cong} \mathrm{Hom}(D^{\#}, C^*)$$

and $g \longmapsto \beta(g)$ describes

$$\mathrm{Hom}(D, C) \xrightarrow{\cong} (C^*, D^{\#})$$

then for $C \xrightarrow{f} D \xrightarrow{g} C'$, $\alpha(f)\beta(g) = (gf)^*$ while for $D \xrightarrow{g} C \xrightarrow{f} D'$, $\beta(g)\alpha(f) = (fg)^{\#}$. Thus corresponding to

$$D \xrightarrow{1} D \xrightarrow{1} D \xrightarrow{1} D$$

we have

$$D^* \xrightarrow{\beta(1)} D^\# \xrightarrow{\alpha(1)} D^* \xrightarrow{\beta(1)} D^\#$$

and $\alpha(1)\beta(1) = 1^* = 1$ while $\beta(1)\alpha(1) = 1\# = 1$ so that $D^* \cong D\#$. Henceforth we can identify D^* and $D\#$ by this isomorphism. Thus $(-)^*$ and $(-)\#$ agree on $\underline{C}\cap\underline{D}$ and we may think of them as determining a single functor

$$(-)^* : (\underline{C} \cup \underline{D})^{op} \longrightarrow \underline{C}\cup\underline{D}$$

that interchanges $\underline{C}$ and $\underline{D}$.

(4.8) If we let $C_2 = I$ in (3.6.iii), we get

$$(C_1,(I,C_3^*)) \cong (C_3,(I,C_1^*)) \ .$$

or, in view of (3.6.i), $(C_1,C_3^*) \cong (C_3,C_1^*)$. If we let $C_3 = I$, we get $(C_1,I^*) \cong C_1^*$. If we let $I^* = T$, this may be summarized as,

Proposition. For $C,C'\epsilon\underline{C}$,

$$C^* \cong (C,T)$$

$$(C',C^*) \cong (C,C'^*)$$

(4.9) Proposition. Let $A,B\epsilon\underline{C}\cup\underline{D}$. Then

$$\text{Hom}(A,B) \cong \text{Hom}(B^*,A^*)$$

Proof. If $A,B\epsilon\underline{C}$ or $A,B\epsilon\underline{D}$ this is just the duality while the other two follow as observed in 4.7.

The naturality of these morphisms is clear, one of the unstated coherence hypotheses is that for $f\epsilon\text{Hom}(A,B), f \longmapsto f^* : B^* \rightarrow A^*$ is an involution.

(4.10) We wish to discuss a condition under which 4.6(iv) follows from the remaining hypotheses. Suppose for every object $C\epsilon\underline{C}$ there is a family of maps $\{m_\psi : C \rightarrow D_\psi\}$, $D_\psi\epsilon\underline{D}$ and for every $D\epsilon\underline{D}$ a family $\{e_\omega : C_\omega \rightarrow D\}$, $C_\omega\epsilon\underline{C}$ which collectively have the properties of a factorization system. Namely suppose that every pair of families $\{f_\omega : C_\omega \rightarrow C\}$ and $\{g_\psi : D \rightarrow D_\psi\}$ such that every diagram

$$\begin{array}{ccc} C_\omega & \xrightarrow{e_\omega} & D \\ {\scriptstyle f_\omega}\downarrow & & \downarrow{\scriptstyle g_\psi} \\ C & \xrightarrow{m_\psi} & D_\psi \end{array}$$

commutes determines a unique map $h:D \rightarrow C$ such that $m_\psi h = g_\psi$ for all ψ and $h\cdot e_\omega = f_\omega$ for all ω.

(4.11) Supposing this to be the case, let a suitable family $\{C \rightarrow D_\psi\}$ be called a $\underline{D}$-representation of C and the dual notion $\{C_\omega \rightarrow D\}$ a $\underline{C}$-generation of D . If $\{C \rightarrow D_\psi\}$ is a $\underline{D}$-representation so is any larger family. In particular so is the family $\{C \rightarrow D_\psi\} \cup \{C \rightarrow C_\omega^*\}$, where $\{C_\omega \rightarrow C^*\}$ is a $\underline{C}$-generation. Thus we may suppose every C has a $\underline{D}$-representation which dualizes to a $\underline{C}$-generation. Similarly every $D\epsilon\underline{D}$ has a $\underline{C}$-generation which dualizes to a $\underline{D}$-representation. Now fixing a map $h:D \rightarrow C$, a D-representation of C and a $\underline{C}$-generation of C both assumed to dualize properly we begin with the diagram

$$\begin{array}{ccc} C_\omega & \xrightarrow{e_\omega} & D \\ f_\omega \downarrow & & \downarrow g_\psi \\ C & \xrightarrow{m_\psi} & D_\psi \end{array}$$

in which f_ω is defined as $h.e_\omega$ and similarly $g_\psi = m_\psi.h$. Dualizing and using 3.9 we get

$$\begin{array}{ccc} D^*_\psi & \xrightarrow{m^*_\psi} & C^* \\ g^*_\psi \downarrow & & \downarrow f^*_\omega \\ D^* & \xrightarrow{e^*_\omega} & C^*_\omega \end{array}$$

which then has a unique fill-in which we naturally denote by h^* . The naturality of $h \mapsto h^*$ and the fact that it is an involution follow readily from the same properties of the C_ω, f_ω, g_ψ and m_ψ .

(4.12) Theorem. Under the hypotheses of 4.6(i),(ii),(iii) and (4.10) we have a duality $\text{Hom}(A,B) \cong \text{Hom}(B^*,A^*)$ for $A,B \in \underline{C} \cup \underline{D}$. Thus we have a pre *-autonomous situation.

(4.13) If we suppose that $\underline{A}$ is a $\underline{V}$ category, we can ask that $(-)^* : \underline{C} \to \underline{D}$ be a $\underline{V}$-functor and that the equivalence be that of $\underline{V}$-enriched homs,

$$\underline{V}(C',C) \cong \underline{V}(C^*,C'^*) \ .$$

In addition, we can suppose that the isomorphism of (4.6.ii) be $\underline{V}$-enriched,

$$\underline{V}(I,(C,D)) \cong \underline{V}(C,D) \ .$$

In that case, we come to the notion of a V-enriched pre-*-autonomous situation. In that case the maps used in the $\underline{C}$-generation and $\underline{D}$-representation in the paragraphs may, if necessary, be replaced by pseudomaps.

(4.13) The $\underline{V}$-enriched versions of (4.8) and (4.9) go through without change. In order to derive the analogue of (4.12) we must modify (4.10). This is done by again supposing a family $\{C_\omega \to D\}$ and $\{C \to D_\psi\}$. We observe that there is, for all ω,ψ, a commutative square,

$$\begin{array}{ccc} \underline{V}(D,C) & \longrightarrow & \underline{V}(D,D_\psi) \\ \downarrow & & \downarrow \\ \underline{V}(C_\omega,C) & \longrightarrow & \underline{V}(C_\omega,D_\psi) \end{array} \ .$$

We may now require that $\underline{V}(D,C)$ be the simultaneous equalizer of all those squares. That is, given any object V and commutative squares

$$\begin{array}{ccc} V & \longrightarrow & \underline{V}(D,D_\psi) \\ \downarrow & & \downarrow \\ \underline{V}(C_\omega,C) & \longrightarrow & \underline{V}(C_\omega,D_\psi) \end{array}$$

one for each ω,ψ, there should exist a unique map $V \to \underline{V}(D,C)$ inducing those squares. Again argumenting the representing and generating families does not change the situation so we may suppose they are invariant under dualization. Then $\underline{V}(D,D_\psi) \cong \underline{V}(D^*_\psi,D^*)$, $\underline{V}(C_\omega,C) \cong \underline{V}(C^*,C^*_\psi)$ and $\underline{V}(C_\omega,D_\psi) \cong \underline{V}(D^*_\psi,C^*_\omega)$ implies that $\underline{V}(C^*,D^*)$ has the same universal mapping property as $\underline{V}(D,C)$ and hence they are isomorphic.

(4.14) Example. On the category of vector spaces, there are at least two reasonable pre-*-autonomous situations. The first is to take $\underline{C} = \underline{D}$ = finite dimensional vector spaces with the usual duality. This is, in fact, already a *-autonomous category but there is no reason not to extend the structure to a larger category. The second is to take $\underline{D}$ to be the category $\underline{V}$, all spaces considered as having the discrete uniformity. For $\underline{C}$ -which must be equivalent to $\underline{D}^{op}$- we take the category of linearly compact spaces. A linearly compact space is a vector space which is, first, topologized linearly. That means that the open sub (vector) spaces form a topological base at 0 . Since the quotient modulo such a subspace is discrete (when you indentify an open set to a point, that point becomes open), this is the same as saying it is a subspace of a product of discrete spaces. A linearly topologized space is linearly compact if every collection of closed linear subvarities (a linear subvariety in V is a set of the form $v + V'$ where v is a point and V' a subspace of V) with the finite intersection property has a non-empty intersection. Lefschetz defined the notion and proved all the elementary properties. Linearly compact spaces are closed under products, separated quotients and closed subspaces. A continuous linear transformation from a linearly compact to a separated space is closed. Lefschetz also showed that every such space is isomorphic (topologically) to a space of the form K^S for a set S . Here since K is linearly compact, so is K^S . As well he showed that the continuous linear maps $K^S \to K^T$ are naturally equivalent to the linear $T \cdot K \to S \cdot K$ which is the statement of the duality. Here $S \cdot K$ and $T \cdot K$ stand for the direct sum of an S-fold, respectively a T-fold of copies of K . See [Lefschetz] pp.78-82 for details. To get the required $(-,-): \underline{C}^{op} \times \underline{D} \to \underline{D}$ simply take the set of continuing linear maps with the usual vector space structure and, of course, the discrete topology.

One observation that may be helpful in thinking about linearly compact spaces is that if K is finite then linear compactness is equivalent to ordinary topological compactness. Then linear compactness may be tought of as the transfer of the notion of compactness from finite fields to arbitrary ones.

CHAPTER II. EXTENSIONS OF STRUCTURE

1. The Setting.

(1.1) The main goal in these notes is to convert a pre-$*$-autonomous situation into a $*$-autonomous category. That is, given a $\underline{V}$-category $\underline{A}$ equipped with subcategories $\underline{C}$ and $\underline{D}$ which determine an enriched pre-$*$-autonomous situation, we wish to find a full subcategory $\underline{G} \subset \underline{A}$ which contains $\underline{C}$ and $\underline{D}$ and can be equipped with a $*$-autonomous structure extending the given structure on $\underline{C}$ and $\underline{D}$.

(1.2) The right degree of abstraction has probably not been reached here. For the purpose of extending the structure on $\underline{C}$ and $\underline{D}$, I have found it expedient to suppose that $\underline{A}$ is a category of uniform objects in a closed category $\underline{V}$ which is a semi-variety.

In this chapter, then, $\underline{V}$ is such a category, Un$\underline{V}$ the category of uniform $\underline{V}$ objects, $\underline{C}$ and $\underline{D}$ are full subcategories equipped with a $\underline{V}$-enriched duality $(-)^*:\underline{C}^{op} \to \underline{D}$, an object $I \in \underline{C}$ and a functor $(-,-):\underline{C}^{op} \times \underline{D} \to \underline{D}$ which give a $\underline{V}$-enriched pre-$*$-autonomous situation. We made the following additional hypotheses: That every object of $\underline{C}$ and for every $V \in \underline{V}$, $D \in \underline{D}$ every object of the form $[V,D]$ can be embedded in a product of objects in $\underline{D}$. We now let $\underline{A}$ denote the full subcategory of Un$\underline{V}$ consisting of all uniform $\underline{V}$-objects which can be embedded in a product of objects of $\underline{D}$. Given $A \in \underline{A}$, a family $\{A \to D_\psi\}$ of maps - or even of pseudomaps - with each $D_\psi \in \underline{D}$ is called a $\underline{D}$-representation of A.

(1.3) Proposition. The inclusion of $\underline{A} \to \text{Un}\underline{V}$ has a left adjoint.
Proof. It is clear from the definition of $\underline{A}$ as the full subcategory of objects which have an embedding into a product in $\underline{D}$ that $\underline{A}$ is itself invariant under products and subobjects and hence under limits. Thus we need only verify the solution set condition. But if we have $U \in \text{Un}\underline{V}$, $A \in \underline{A}$ and a map $U \to A$, the image has cardinality $\le$ that of U and moreover lies in $\underline{A}$. Thus all the algebras of $\underline{A}$ whose cardinality do not exceed that of U and all possible maps of U to such algebras constitute a solution set.

(1.4) For $A \in \underline{A}$ consider the diagram

$$\begin{array}{ccc}
\underline{A}^{op} & \underset{\check\sigma}{\overset{\sigma}{\rightleftarrows}} & (\text{Un}\underline{V})^{op} \\
\underline{V}(-,A)\Big\downarrow & [-,A]\ \Big\uparrow\ \Big\uparrow & \underline{V}(-,A) \\
\underline{V} & \underset{\check\tau}{\overset{\tau}{\rightleftarrows}} & \underline{V}
\end{array}$$

with σ inclusion, $\check\sigma$ the adjoint above and τ and $\check\tau$ are the identity. With the variance exhibited, $[-,A]$ is left adjoint to $\underline{V}(-,A)$. The adjunction $\sigma \dashv \check\sigma$ gives $\underline{V}(U,A) \cong \underline{V}(\check\sigma U,A)$ which is the commutativity required to apply [Barr,73], Theorem 3 and conclude that the image of $[-,A]$ lies in $\underline{A}$. This is clearly the cotensor. Thus $\underline{A}$ is a cotensored $\underline{V}$-category.

(1.5) It is easy to see that we get a factorization system on $\underline{A}$ by taking $\underline{E}$ to be the surjections and $\underline{M}$ the embeddings (with the induced uniformity). Suppose now we have a surjection $A' \longrightarrow A$ and an embedding $B \longrightarrow B'$. We want to claim that

$$\begin{array}{ccc} \underline{V}(A,B) & \longrightarrow & \underline{V}(A',B) \\ \downarrow & & \downarrow \\ \underline{V}(A,B') & \longrightarrow & \underline{V}(A',B') \end{array}$$

is a pullback. This is so iff for every $V \in \underline{V}$ the functor $\mathrm{Hom}(V,-)$ applied to the above square is a pullback in $\underline{S}$. Using the cotensor adjointness, we get a commutative square

$$\begin{array}{ccc} \mathrm{Hom}(A,[V,B]) & \longrightarrow & \mathrm{Hom}(A',[V,B]) \\ \downarrow & & \downarrow \\ \mathrm{Hom}(A,[V,B']) & \longrightarrow & \mathrm{Hom}(A',[V,B']) \end{array}$$

which we need to know is a pullback. According to (I.3.13) when B is a subspace of B' , $[V,B]$ is a subspace of $[V,B']$ so that we have a diagonal fill-in in any commutative square

$$\begin{array}{ccc} A' & \longrightarrow & A \\ \downarrow & & \downarrow \\ [V,B] & \longrightarrow & [V,B'] \end{array}$$

which is exactly what is required. Thus the category $\underline{A}$, the subcategories $\underline{C}$ and $\underline{D}$ and the remaining structure fulfill all the conditions of I.4.

(1.6) Example. Let $\underline{V}$ be the category of vector spaces over the field K . We consider two possibilities for the pair $\underline{C}$, $\underline{D}$. For the first we take $\underline{C} = \underline{D}$ = finite dimensional vector spaces (with the discrete topology). Then every $C \in \underline{C}$ is a power of K and hence has a $\underline{D}$-representation. Since the spaces in $\underline{D}$ have linear topologies so does anything with a $\underline{D}$-representation. Not every linearly topologized space has a $\underline{D}$-representation however. In fact, if A is embedded in ΠD_ω and every $D_\omega \in \underline{D}$ then every open subspace contains a finite intersection of the kernels of maps $A \rightarrow D_\omega$, that is, the kernel of $A \rightarrow D_{\omega 1} \times \ldots \times D_{\omega n}$. The latter product is finite dimensional, so that the kernel — and with it every open subspace — is cofinite dimensional. In particular this is not true of any infinite dimensional discrete space and no such space belongs to $\underline{A}$. It is easy to see that a space does have a $\underline{D}$-representation iff it has the following property which is a linear analogue to the property of a uniform space being totally bounded. Namely, we say that a space A is linearly totally bounded iff for every open subspace $B \subset A$, there is a finite number of elements $a_1, \ldots, a_n$ such that A is in the linear span of B and $a_1, \ldots, a_n$. Of course this is just a rewording of the hypothesis that A/B is finite dimensional. The analogy is hightened by the following fact.

Proposition. A separated linearly topologized space is linearly totally bounded iff its uniform completion is linearly compact.

Proof. If A is linearly totally bounded, then every open subspace $A_\omega \subset A$ is cofinite dimensional. Thus A/A_ω is finite dimensional. The maps $A \to A/A_\omega$ combine to give a map $A \to \Pi A/A_\omega$ which I claim is an embedding. In fact its kernel is $\cap A_\omega = 0$ because A is separated. Any neighborhood of 0 contains an open sub(vector) space A_ψ which is the inverse image of the subset of $\Pi A/A_\omega$ consisting of 0 in the ψ coordinate and A/A_ω in all others. This is evidently open and shows that A is embedded in the product. The closure of A is then a closed subspace of a product of linearly compact spaces and hence is linearly compact.

To see the converse first suppose that A is linearly compact. Then for any open subspace $A_0 \subset A$, A/A_0 is discrete and also linearly compact and hence finite dimensional (see [Lefschetz]). Now if $B \subset A$ any neighborhood U of 0 in B is the form $U = B \cap V$ where V is a neighborhood of 0 in A . Then V contains an open subspace A_0 whence $U \supset B_0 = B \cap A_0$. This shows that B is linearly topologized and also that B/B_0 is finite dimensional since it maps injectively to A/A_0 .

The other choice for $\underline{C}$ and $\underline{D}$ is to take $\underline{D}$ to be the discrete spaces and $\underline{C}$ the linearly compact ones. Then a space has a $\underline{D}$-representation iff it is linearly topologized. For as noted earlier that is equivalent to being a subspace of a product of discrete spaces, i.e. to having a $\underline{D}$-representation. Thus in that case $\underline{A}$ consists of the linearly topologized spaces.

2. Extension of the Duality.

(2.1) The first task is to extend the duality on $\underline{C} \cup \underline{D}$ to all of $\underline{A}$ or at least to a large full subcategory. Let $A \in \underline{A}$. We are trying to define a dual of A , which we temporarily designate $A^\#$. If we have a map $C \to A$ there should be one $A^\# \to C^*$. If we wish to have the duality be $\underline{V}$-enriched this means $\underline{V}(C,A) \to \underline{V}(A^\#,C^*)$ which means that corresponding to each pseudomap $C \to A$ we require a pseudomap $A^\# \to C^*$. Moreover, we will eventually want to extend the internal hom as well in such a way that the dual of A is its internal hom into T . This requires that $|A^\#| \cong \underline{V}(A,T)$. Accordingly we define $A^\#$ to be the object $\underline{V}(A,T)$ equipped with the coarsest uniformity such that corresponding to each $C \in \underline{C}$ and each element $\underline{A}(C,A)$, the corresponding element of $\underline{V}(|A^\#|, |C^*|)$ under $\underline{V}(C,A) \longrightarrow \underline{V}(\underline{V}(A,T),\underline{V}(C,T))$ is uniform. Another way to describe this uniformity is to say that we equip $\underline{V}(A,T)$ with the coarsest uniformity such that

$$A^\# \longrightarrow [\underline{V}(C,A),C^*]$$

is uniform for all $C \in \underline{C}$. Since $A^\#$ can have only a set of uniform covers, only a set of C need be used from any one A . By taking C = I (which may always be included among the objects of $\underline{C}$ used) we get

$$A^{\#} \longrightarrow [\underline{V}(I,A), I^*] = [\,|A|,\ T]$$

among these maps. This map is an injection since the underlying map in $\underline{V}$ factors as injections

$$|A^{\#}| = \underline{V}(A,T) \longrightarrow \underline{V}(|A|,|T|) \cong |[\,|A|\,,T]\,| \;.$$

Thus there is a family $\{C_{\omega}\}$ such that $A^{\#}$ is embedded in a product $\Pi[\underline{V}(C_{\omega},A),C^*_{\omega}]$.

(2.2) This means that to map $B \longrightarrow A^{\#}$, we require two things. First, we need a map $|B| \longrightarrow \underline{V}(A,T)$. Second we need $B \longrightarrow [\,\underline{V}(C,A),C^*]$, for each $C \in \underline{C}$ such that

$$\begin{array}{ccc} |B| & \longrightarrow & \underline{V}(A,T) \\ \downarrow & & \downarrow \\ |[\underline{V}(C,A),C^*]| & \cong & \underline{V}(\underline{V}(C,A),\underline{V}(C,T)) \end{array}$$

commutes, the right hand map being composition. This is equivalent to maps $\underline{V}(C,A) \to \underline{V}(B,C^*)$ such that

$$\begin{array}{ccc} \underline{V}(C,A) & \longrightarrow & \underline{V}(B,C^*) \\ & \searrow & \downarrow \\ & & \underline{V}(\underline{V}(A,T),\underline{V}(C,T)) \end{array}$$

commutes.

(2.3) Thus to map $C^* \to C^{\#}$ for $C \in \underline{C}$ we need first $\langle C^* \rangle \longrightarrow \langle C^{\#} \rangle$. But each is isomorphic to $\underline{V}(C,T)$ so we can take the composite of the canonical isomorphisms. The second datum required is a map $\underline{V}(C',C) \longrightarrow \underline{V}(C^*,C'^*)$ for which we take the duality isomorphism. The required coherence is trivial. On the other hand among the candidates for C_{ω} we may take C itself so that the identity gives a map

$$C^{\#} \longrightarrow [\,\underline{V}(C,C),C^*]$$

which composed with the unit $I \to \underline{V}(C,C)$ gives $C^{\#} \to C^*$. Thus $C^* \cong C^{\#}$.

(2.4) Now let $D \in \underline{D}$. We know that there is a $\underline{C}$-generating family $\{C_{\omega} \longrightarrow D\}$ such that D^* is isomorphic to a subobject of ΠC^*_{ω} . Since $D^{\#} \longrightarrow \Pi\,[\,\underline{V}(C_{\omega},D),C^*_{\omega}]$ is a map, we may follow each with the name $I \longrightarrow \underline{V}(C_{\omega},D)$ of the corresponding maps $C_{\omega} \longrightarrow D$ to get $D^{\#} \longrightarrow \Pi C^*_{\omega}$, and hence $D^{\#} \longrightarrow D^*$. To go the other way, we require for all $C \in \underline{C}$

$$\underline{V}(C,D^*) \longrightarrow \underline{V}(D,C^*)$$

for which we take the duality isomorphism.

We have now established that $(\ )^{\#}$ is an extension of $(\ )^{\#}$ and will henceforth write A^* instead of $A^{\#}$. We now turn to the functionality of this operation.

(2.5) Proposition. For any $C \in \underline{C}$, composition of maps gives a map $\underline{V}(C,A) \longrightarrow \underline{V}(A^*,C^*)$.
Proof. This is the same as a map

$$A^* \longrightarrow [\underline{V}(C,A),C^*]$$

which we have.

Corollary. For any $A,B \in \underline{A}$, composition of maps gives a map $\underline{V}(B,A) \longrightarrow \underline{V}(A^*,B^*)$.
Proof. We require a map

$$A^* \longrightarrow [\underline{V}(B,A),B^*] .$$

We certainly have a map

$$\underline{V}(A,T) \longrightarrow \underline{V}(\underline{V}(B,A),\underline{V}(B,T)) .$$

Moreover, for any $C \in \underline{C}$ we have

$$\begin{aligned} A^* &\longrightarrow [\underline{V}(C,A),C^*] \\ &\longrightarrow [\underline{V}(B,A) \otimes \underline{V}(C,B),C^*] \\ &\cong [\underline{V}(B,A),[\underline{V}(C,B),C^*]] \end{aligned}$$

and hence there is the required map $A^* \longrightarrow [\underline{V}(B,A),B^*]$. Note that this argument uses the fact that $[V,-]$ preserves embeddings.

(2.6) This shows that $(-)^* : \underline{A} \longrightarrow \underline{A}$ is a $\underline{V}$-functor. There is nothing to guarantee that it is an equivalence and it is in fact unlikely that it is always so.

Since $I \in \underline{C}$, composition gives a map

$$A^* \longrightarrow [\underline{V}(I,A),T]$$
$$|A| = \underline{V}(I,A) \longrightarrow \underline{V}(A^*,T) = |A^{**}|$$

and the best you can usually hope for is that the above map $|A| \longrightarrow |A^{**}|$ be an isomorphism. It is always a monomorphism. For there is a $\underline{D}$-representation $\{A \longrightarrow D_\omega\}$ which gives $\{A^{**} \longrightarrow D^{**}_\omega \cong D_\omega\}$. Thus we have $|A| \longrightarrow |A^{**}| \longrightarrow \Pi|D_\omega|$ is a monomorphism and hence the first map is.

(2.7) Let A^Δ denote the subobject of A^{**} such that $|A^\Delta|$ is the image of $|A|$ in $|A^{**}|$ but such that A^Δ has the induced uniformity. There is no question of $A^\Delta \in \underline{A}$ since $A^\Delta \subset A^{**}$. We have $|A^\Delta| \longrightarrow |A|$ by the inverse of the above inclusion. Let $\{A \longrightarrow D_\omega\}$ be a $\underline{D}$-representation of A . Then each $A \longrightarrow D_\omega$ gives $A^{**} \longrightarrow D^{**}_\omega \cong D_\omega$ so that we have $A^\Delta \longrightarrow A$ is uniform. We say that A is prereflexive if $A^\Delta \longrightarrow A$ is an isomorphism, quasireflexive if $A^\Delta = A^{**}$ and reflexive if both of these hold.

(2.8) The condition that A be quasireflexive is equivalent to the assertion that

$$\underline{V}(I,A) \cong \underline{V}(A^*,I^*) .$$

or that every pseudomap $A^* \longrightarrow T$ is represented by evaluation at an element — necessarily unique — of A .

(2.9) Proposition. Suppose

(i) T is cosmall in $\underline{A}$ (see proof);

(ii) T is injective in the $\underline{V}$-category $\underline{A}$ with respect to the class of embeddings;

(iii) $\underline{C}$ is closed under finite sums and $\underline{D}$ under finite products and these have the universal mapping properties for pseudomaps as well as maps.

Then every pseudomap $f : A^* \longrightarrow T$ is represented by evaluation at an element of A .

Proof. We can find a family of pseudomaps $\{C_\omega \to A\}$ such that the horizontal arrow in the diagram

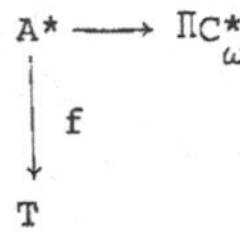

is an embedding. Then we have a map $f^\# : \Pi C_\omega^* \longrightarrow T$ which extends f . The hypothesis that T be cosmall means that such a map factors through a product of finitely many, say $C_1^* \times \ldots \times C_n^*$. Since $\underline{D}$ is closed under finite products this is also their product in $\underline{D}$. Since $C_1 + C_2 + \ldots + C_n \in \underline{C}$, that is their sum in $\underline{C}$. Since $(-)^* : \underline{C}^{op} \longrightarrow \underline{D}$ is an equivalence, $(C_1 + \ldots + C_n)^* \cong C_1^* \times \ldots \times C_n^*$ so we have

$$\begin{array}{ccc} A^* & \longrightarrow & \Pi C_\omega^* \\ \downarrow & & \downarrow \\ T & \longleftarrow & (C_1+\ldots+C_n)^* \end{array} .$$

From duality for $\underline{C}$ it follows that the lower pseudomap is represented by an $x \in C_1 + \ldots + C_n$. The image of that element under the pseudomap $C_1 + \ldots + C_n \longrightarrow A$ whose components are the ones given originally is easily seen to represent f .

Further discussion of the duality is postponed till the next section.

(2.10) Example. In the example of vector spaces, the hypotheses of (2.10) are satisfied for both possible choices of $\underline{C}$ and $\underline{D}$, in fact for any choice for which the spaces are linearly topologized. For if B is embedded in A and $B \longrightarrow K$ is a continous linear map, its kernel must be an open subspace $B_0 \subset B$. Then as we showed in 1.6, $B_0 = A_0 \cap B$ where A_0 is an open subspace in A . Since $B/B_0 \longrightarrow A/A_0$ is an injection and both are discrete, the former is embedded in the latter. Then we have a diagram

$$\begin{array}{ccccc} B & \longrightarrow & B/B_0 & \longrightarrow & K \\ \downarrow & & \downarrow & & \\ A & \longrightarrow & A/A_0 & & \end{array}$$

and then ordinary vector space theory provides the required $A/A_0 \longrightarrow K$, continous since A/A_0 is discrete.

3. Extension of the Internal Hom

(3.1) As with the duality, we wish to extend the functor $(-,-) : \underline{C}^{op} \times \underline{D} \to \underline{D}$ to a functor denoted

$$\underline{A}(-,-) : \underline{A}^{op} \times \underline{A} \to \underline{A} .$$

Although we are for convenience using notation suggesting this is an internal hom, it is not in general. It is not generally symmetric, does not always have an adjoint and when it does, the tensor is not always associative. Chapter III is devoted to finding a nice subcategory on which it is well-behaved.

(3.2) To begin with we require $|\underline{A}(A,B)| = \underline{V}(A,B)$, the $\underline{V}$-valued hom. Second we will give it the coarsest uniformity such that the pseudomap $\underline{A}(A,B) \to (C,D)$ is uniform for every pseudomap $C \to A$ and every pseudomap $B \to D$. More abstractly, we require that $\underline{A}(A,B)$ have the weak uniformity determined by all $C \in \underline{C}$, $D \in \underline{D}$ and $\underline{A}(A,B) \longrightarrow [\underline{V}(C,A) \otimes \underline{V}(B,D), (C,D)]$. This makes sense for the underlying map in $\underline{V}$ is

$$\underline{V}(A,B) \longrightarrow \underline{V}(\underline{V}(C,A) \otimes \underline{V}(B,D), \underline{V}(C,D))$$

which is the transpose under adjunction to composition

$$\underline{V}(C,A) \otimes \underline{V}(A,B) \otimes \underline{V}(B,D) \longrightarrow \underline{V}(C,D) \quad .$$

Since there is an epimorphic family $\{f_\omega : C_\omega \longrightarrow A\}$ and a monomorphic family $\{g_\psi : B \to D_\psi\}$, there is a monomorphism

$$\underline{V}(A,B) \longrightarrow \Pi \underline{V}(C_\omega, D_\psi) .$$

This map factors

$$\underline{V}(A,B) \longrightarrow \Pi \underline{V}(\underline{V}(C_\omega,A) \otimes \underline{V}(B,D_\psi), \underline{V}(C_\omega,D_\psi)) \to \Pi \underline{V}(I \otimes I, \underline{V}(C_\omega,D_\psi)) \cong \Pi \underline{V}(C_\omega,D_\psi)$$

where the second map is induced by the names of the f_ω and g_ψ. Thus the first map is an injection as well. Hence for some (not necessarily the same) families $\{C_\omega\}$ and $\{D_\psi\}$ of objects of $\underline{C}$ and $\underline{D}$ respectively, $\underline{A}(A,B)$ is embedded in

$$\Pi [\underline{V}(C_\omega,A) \otimes \underline{V}(B,D_\psi), (C_\omega,D_\psi)] \quad .$$

(3.3) Proposition. If $C \in \underline{C}$ and $D \in \underline{D}$, $\underline{A}(C,D)$ is canonically isomorphic to (C,D).
Proof. Since $(-,-) : \underline{C}^{op} \times \underline{D} \to \underline{D}$ is assumed to be a $\underline{V}$-functor, there is for each $D \in \underline{D}$, a natural map

$$\underline{V}(C',C) \to \underline{V}((C,D),(C',D))$$

which expresses the fact that $(-,D)$ is a $\underline{V}$-functor. Similarly, there is for each $C \in \underline{C}$, a map

$$\underline{V}(D,D') \to \underline{V}((C,D),(C,D')) \; .$$

Each of these maps lies alone ordinary composition of functions. Putting these together and replacing C by C' in the second we get a map

$$\underline{V}(C',C) \otimes \underline{V}(D,D') \to \underline{V}((C,D),(C',D)) \otimes \underline{V}((C',D),(C',D')) \to \underline{V}((C,D),(C',D')) \; ,$$

the second map being composition. By the cotensor adjunction, this gives a map,

$$(C,D) \to [\underline{V}(C',C) \otimes \underline{V}(D,D'), (C',D')]$$

for any $C' \in \underline{C}$, $D' \in \underline{D}$. This implies that $(C,D) \to \underline{A}(C,D)$ is uniform. To go the other way we observe that

$$\underline{A}(C,D) \to [\underline{V}(C,C) \otimes \underline{V}(D,D), (C,D)]$$

must be uniform. We may compose this with the map induced by names of the identity maps of C and D respectively to get $\underline{A}(C,D) \to [I \otimes I, (C,D)] \cong (C,D)$, the latter isomorphism coming from

$$\underline{V}(A,[I,B]) \cong \underline{V}(I,\underline{V}(A,B)) \cong \underline{V}(A,B)$$

from which $[I,B] \cong B$ by the Yoneda lemma. This implies that $\underline{A}(C,D) \to (C,D)$ is uniform. It is easy to see that both of these maps lie over the identity map on $\underline{V}(C,D)$.

(3.4) Proposition. The bifunctor $\underline{A}(-,-) : \underline{A}^{op} \times \underline{A} \to \underline{A}$ is (or lifts to) a $\underline{V}$-functor.
Proof. We must show that there are natural maps

$$\underline{V}(A',A) \to \underline{V}(\underline{A}(A,B), \underline{A}(A',B))$$

$$\underline{V}(B,B') \to \underline{V}(\underline{A}(A,B), \underline{A}(A,B'))$$

for all $A, A', A, B' \in \underline{A}$. We do the first, the second being similar. Then we require a map

$$\underline{A}(A,B) \to [\underline{V}(A',A), \underline{A}(A',B)]$$

which means first a map $\underline{V}(A,B) \to \underline{V}(\underline{V}(A',A), \underline{V}(A',B))$ and second for all $C \in \underline{C}$, $D \in \underline{D}$,

$$\underline{A}(A,B) \to [\underline{V}(A',A), [\underline{V}(C,A') \otimes \underline{V}(B,D), (C,D)]]$$

by (I.3.13). The first is just composition and the second comes from

$$\begin{aligned} \underline{A}(A,B) &\to [\underline{V}(C,A) \otimes \underline{V}(B,D), (C,D)] \\ &\to [\underline{V}(A',A) \otimes \underline{V}(C,A') \otimes \underline{V}(B,D), (C,D)] \\ &\cong [\underline{V}(A',A), [\underline{V}(C,A') \otimes \underline{V}(B,D), (C,D)]] . \end{aligned}$$

(3.5) Proposition. Suppose $\underline{A}(A,B)$ is embedded in the product $\Pi[\underline{V}(C_\omega,A) \otimes \underline{V}(B,D_\psi), (C_\omega,D_\psi)]$. Then there is a commutative diagram

$$\begin{array}{ccc} \underline{A}(A,B) & \longrightarrow & \prod_\omega [\underline{V}(C_\omega,A), (C_\omega,B)] \\ \downarrow & & \downarrow \\ \prod_\psi [\underline{V}(B,D_\psi), (B,D_\psi)] & \longrightarrow & \prod_{\omega,\psi} [\underline{V}(C_\omega,A) \otimes \underline{V}(B,D_\psi), (C_\omega,D_\psi)] \end{array} .$$

The upper map and left hand map are embeddings.
Proof. From (3.4) we have a map for all ω,

$$\underline{V}(C_\omega,A) \to \underline{V}(\underline{A}(A,B), \underline{A}(C_\omega,B))$$

which transposes to

$$\underline{A}(A,B) \longrightarrow [\underline{V}(C_\omega,A),\underline{A}(C_\omega,B)]$$

and is the ω component of a map

$$\underline{A}(A,B) \rightarrow \prod_\omega [\underline{V}(C_\omega,A),\underline{A}(C_\omega,B)] \ .$$

Similarly, there are maps for all ω,ψ

$$\underline{V}(B,D_\psi) \rightarrow \underline{V}(\underline{A}(C_\omega,B),(C_\omega,D_\psi))$$

which transpose to

$$\underline{A}(C_\omega,B) \rightarrow [\underline{V}(B,D_\psi),(C_\omega,D_\psi)] \ .$$

Cotensoring $\underline{V}(C_\omega,A)$ and using the fact that the cotensor is a $\underline{V}$-functor, we get

$$[\underline{V}(C_\omega,A),\underline{A}(C_\omega,B)] \longrightarrow [\underline{V}(C_\omega,A)\otimes\underline{V}(B,D_\psi),(C_\omega,D_\psi)] \ .$$

This is the ψ component of a map to $\prod_\psi [\underline{V}(C_\omega,A)\otimes\underline{V}(B,D_\psi),(C_\omega,D_\psi)]$. Finally the product over all ω gives a map

$$\prod_\omega [\underline{V}(C_\omega,A),\underline{A}(C_\omega,B)] \rightarrow \prod_{\omega,\psi} [V(C_\omega,A)\otimes\underline{V}(B,D_\psi),(C_\omega,D_\psi)] \ .$$

This gives the one factorization and the other is analogous. The last property follows from cancellation properties of factorization systems.

Corollary 1. If $C\in\underline{C}$, $\underline{A}(C,B)$ is embedded in a product $\prod[\underline{V}(B,D_\psi),(C,D_\psi)]$; if $D\in\underline{D}$, $\underline{A}(A,D)$ is embedded in a product $\prod[\underline{V}(C_\omega,A), (C_\omega,D)]$.

Corollary 2. For any $A\in\underline{A}$, $A^* \cong \underline{A}(A,T)$.

Corollary 3. There is a canonical isomorphism $\underline{A}(I,A) \cong A$.

Proof. By hypothesis, $|A| = \underline{V}(I,A) \cong |(I,A)|$. For some family $\{D_\psi\}$ of objects of $\underline{D}$, A is canonically embedded in $\prod[\underline{V}(A,D_\psi), D_\psi]$ while (I,A) is canonically embedded in the isomorphic $\prod[\underline{V}(A,D_\psi), (I,D_\psi)]$.

Corollary 4. There is a canonical map

$$\underline{A}(A,B) \longrightarrow [\ |A|\ ,\ B] \ .$$

Proof. We have, from (3.4), the canonical map

$$|A| = \underline{V}(I,A) \rightarrow \underline{V}(\underline{A}(A,B),\underline{A}(I,B)) \cong \underline{V}(\underline{A}(A,B),B)$$

which has the transpose

$$\underline{A}(A,B) \longrightarrow [\ |A|\ ,B] \ .$$

(3.6) Proposition. Suppose A and B are reflexive. Then $\underline{A}(A,B) \cong \underline{A}(B^*,A^*)$.

Proof. Since $(-)^*$ is a $\underline{V}$-functor, we have $\underline{V}(A,B) \rightarrow \underline{V}(B^*,A^*) \rightarrow \underline{V}(A^{**},B^{**})$ an equivalence. The first is thus a split mono and the second a split epi. The second is an instance of the first and hence is also a split mono from which it follows that both are isomorphisms. Thus $\underline{V}(A,B) \cong \underline{V}(B^*,A^*)$. In particular we have $\underline{V}(C,A) \cong \underline{V}(A^*,C^*)$ and $\underline{A}(B,D) \cong \underline{A}(D^*,B^*)$ for $C\in\underline{C}$ and $D\in\underline{D}$. Thus we have

$$\begin{aligned}\underline{A}(B^*,A^*) &\rightarrow [\underline{V}(D^*,B^*)\otimes\underline{V}(A^*,C^*),(D^*,C^*)] \\ &\cong [\underline{V}(C,A)\otimes\underline{V}(B,D),(C,D)]\end{aligned}$$

so we have $\underline{A}(B^*,A^*) \rightarrow \underline{A}(A,B)$ and similarly in the other direction. Note that $(-)^*$

is not in general an $\underline{A}$ functor. In fact if B is quasi-reflexive and not reflexive, the map

$$B \cong \underline{A}(I,B) \longrightarrow \underline{A}(B^*,T) = B^{**}$$

is not uniform.

(3.7) Let $C \in \underline{C}$, $D \in \underline{D}$, S a subset of C and d a pseudometric on D. Then (d,S) is the pseudometric defined on $\underline{V}(C,D)$ by

$$(d,S)(f,g) = \sup \{d(fx,gx) \mid x \in S\} .$$

If $\Phi = \Phi(C)$ is a collection of subsets of C then the collection of pseudometrics (d,S), $S \in \Phi$, d a pseudometric on D, defines a pre-(i.e. not necessarily separated) uniformity on C, that of <u>uniform convergence on the sets in</u> Φ. If the union of the sets in Φ is dense in C (e.g. if Φ includes all singletons) then this structure is separated. For if $f \neq g$ there is an element of the dense union, hence an element $x \in S \in \Phi$ such that $fx \neq gx$. Since D is separated there is a pseudometric d for which $d(fx,gx) \neq 0$. We say that the uniform structure on the values of the functor $(-,-) : \underline{C}^{op} \times \underline{D} \to \underline{D}$ is a <u>convergence uniformity</u> if there is given for each $C \in \underline{C}$ a family $\Phi(C)$ of subsets of C such that (C,D) has the structure of uniform convergence on the sets in $\Phi(C)$ and if, moreover, for every pseudomap $f : C \to C'$ and $S \in \Phi(C)$ there is an $S' \in \Phi(C')$ such that $f(S) \subset S'$. In practice, $\Phi(C)$ usually consists of something like all compact sets or all finite sets or the like. The second requirement amounts to supposing that there is a natural transformation

$$|\underline{V}(C,C')| \longrightarrow \mathrm{Hom}(\Phi(C),\Phi(C'))$$

which is something like a $\underline{V}$-enrichment. At any rate it is enough to guarantee that there is a canonical

$$\underline{V}(C,C') \longrightarrow \underline{V}((C',D),(C,D))$$

for any $D \in \underline{D}$. The similar

$$\underline{V}(D,D') \longrightarrow \underline{V}((C,D),(C,D'))$$

for any $C \in \underline{C}$ exists because the inverse image of a pseudometric is a pseudometric.

(3.8) We now let $\Phi(A)$ denote the family of all subsets of A of the form $S_1 \cup \ldots \cup S_n$ where for each $i = 1,\ldots,n$, S_i is the image under same pseudomap $C_i \to A$ of a set in $\Phi(C_i)$. If a set $\{C_\omega \to A\}$ of pseudomaps with $C_\omega \in \underline{C}$ is sufficient to generate $\Phi(A)$ in this manner, we say that the family $\{C_\omega \to A\}$ generates A.

Dually if $\{B \to D_\omega\}$ is a collection of pseudomaps such that B is embedded in ΠD_ω, then we say that the product <u>represents</u> B.

(3.9) <u>Proposition</u>. Let $A,B \in \underline{A}$. Then $\underline{A}(A,B)$ has the uniformity of uniform convergence on $\Phi(A)$. If $\{C_\omega \to A\}$ generates A and $\{B \to D_\psi\}$ represents B then $\{\underline{A}(A,B) \longrightarrow (C_\omega, D_\psi)\}$ represents (A,B).

Proof. Let, momentarily, $\underline{A}\#(A,B)$ denote $\underline{V}(A,B)$ equipped with the uniformity of uniform convergence on $\Phi(A)$. Then a basis of pseudometrics on $\underline{A}\#(A,B)$ consists of

(d,S) , d a pseudometric on A , $S\in\Phi(A)$ (see 3.7). Let $S = S_1\cup\ldots\cup S_n$, $S_i = (\Phi f_i)(T_i)$, $T_i\in\Phi(C_i)$, $f_i\in\underline{V}(C_i,A)$. Also let $d = \sup\{d_j(g_j,g_j)\}$, $g_j\in\underline{V}(B,D_j)$, $j = 1,\ldots,m$. Then it is easily seen that $(d,S) = \sup\{(d_j(g_jf_i,g_jf_i),T_i)\}$, $i = 1,\ldots,n$, $j = 1,\ldots,m$. But this is just the pseudometric induced by the composite function

$$\underline{A}\#(A,B) \to \Pi(C_i,D_j)\ .$$ Thus $A\#(A,B)$ is embedded in $\Pi(C_\omega,D_\psi)$. But the uniformity on $\underline{A}(A,B)$ is such that any pseudomap $\underline{A}(A,B) \to \Pi(C_\omega,D_\psi)$ is uniform when it is induced by elements of $\underline{V}(C_\omega,A)$ and $\underline{V}(B,D_\psi)$. Thus it follows that the map $\underline{A}(A,B) \to \underline{A}\#(A,B)$ is uniform.

To go the other way, let d be a pseudometric on (A,B) . Then there is a finite set of pseudomaps,say $f_i : C_i \to A$, $g_i : B \to D_i$, $i = 1,\ldots,n$, sets $T_i\in\Phi(A_i)$ and pseudometrics d_i on D_i such that $d \le \sup\{(d_i(g_if_i,g_if_i),T_i)\}$, $i = 1,\ldots,n$. But $\cup\Phi(f_i)T_i\in\Phi(A)$ and is thus already realized by a finite number of elements of the sets $\underline{V}(C_\omega,A)$ and sets from $\Phi(C_\omega)$. Similarly, the pseudometric $\sup(d_i(g_i,g_i))$ on B must be majorized by the sup of pseudometrics arising from a finite number of pseudomaps $B \to D_\psi$. Thus every pseudometric on $\underline{A}(A,B)$ is a pseudometric on $\underline{A}\#(A,B)$ and so $\underline{A}\#(A,B) \to \underline{A}(A,B)$ is uniform.

(3.10) We say that a map $A'\to A$ is <u>dominating</u> provided $\Phi(A')$ maps onto $\Phi(A)$. It is clear that this is dual to an embedding. A family $\{C_\omega\to A\}$ generates A iff $\Sigma C_\omega \to A$ dominates just as $\{B \to D_\psi\}$ represents B iff $B \to \Pi D_\psi$ is an embedding.

<u>Proposition</u>. Suppose $A'\to A$ is dominating and $B \to B'$ is an embedding. Then each map in the square

$$\begin{array}{ccc} \underline{A}(A,B) & \longrightarrow & \underline{A}(A',B) \\ \downarrow & & \downarrow \\ \underline{A}(A,B') & \longrightarrow & A(A',B') \end{array}$$

is an embedding.

Proof. Just repeat the previous argument. No special property of $\underline{C}$ or $\underline{D}$ was used in that argument.

<u>Corollary</u>. Suppose $\{A_\omega\to A\}$ and $\{B \to B_\psi\}$ are families such that $\Sigma A_\omega\to A$ is dominating and $B \to \Pi B_\omega$ is an embedding. Then each map in the square

$$\begin{array}{ccc} \underline{A}(A,B) & \longrightarrow & \Pi\underline{A}(A_\omega,B) \\ \downarrow & & \downarrow \\ \Pi\underline{A}(A,B_\psi) & \longrightarrow & \Pi\underline{A}(A_\omega,B_\psi) \end{array}$$

is an embedding.

(3.11) <u>Proposition</u>. Let $\{C_\omega\to A\}$ and $\{C_\psi\to B\}$ dominate A and B , respectively, and $\{E \to D_\xi\}$ represent E . Then $\underline{A}(A,\underline{A}(B,E))$ is embedded in $\Pi(C_\omega, (C_\psi,D_\xi))$.

Proof. We know from (3.9) that $\underline{A}(B,E)$ is embedded in $\Pi(C_\psi,D_\xi)$ so that $\{\underline{A}(B,E)\to (C_\psi,D_\xi)\}$ represents it. A second application of (3.9) yields the result.

We note that this implies that $\underline{A}(A,\underline{A}(B,E))$ is embedded in

$$\Pi[\underline{A}(C_\omega,A)\otimes\underline{A}(C_\psi,B)\otimes\underline{A}(E,D_\xi),(C_\omega,(C_\psi,D_\xi))]\ .$$

By symmetry so is $\underline{A}(B,\underline{A}(A,E))$. This does not imply that they are isomorphic but it does imply that any inclusion between $\underline{V}(A,\underline{A}(B,E))$ and $\underline{V}(B,\underline{A}(A,E))$ (each considered as a subobject of $\underline{V}(|A|\otimes|B|,|E|)$ under the isomorphism with $\underline{V}(|B|\otimes|A|,|E|)$) automatically lifts to an inclusion between $\underline{A}(A,\underline{A}(B,E))$ and $\underline{A}(B,\underline{A}(A,E))$. The best we can do now is (3.13). When we introduce completeness hypotheses in the next chapter, we will have better results.

(3.12) Proposition. Suppose $A,B,B' \in \underline{A}$ and $B \longrightarrow B'$ is an embedding. Then there is a canonical pullback

$$\begin{array}{ccc} \underline{V}(A,B) & \dashrightarrow & \underline{V}(A,B') \\ \downarrow & & \downarrow \\ \underline{V}(|A|,|B|) & \dashrightarrow & \underline{V}(|A|,|B'|) \end{array}$$

Proof. An element of the pullback is a pseudomap $A \longrightarrow B'$ for which there is a factorization $|A| \longrightarrow |B| \longrightarrow |B'|$. Since B is embedded in B' , the map $|A| \rightarrow |B|$ underlies a uniform $A \longrightarrow B$.

Corollary 1. There is a canonical pullback

$$\begin{array}{ccc} \underline{A}(A,B) & \longrightarrow & \underline{A}(A,B') \\ \downarrow & & \downarrow \\ [\,|A|,B] & \dashrightarrow & [\,|A|,B'] \end{array}$$

Proof. The diagram in $\underline{V}$ which underlies this is a pullback so the actual pullback has the same underlying $\underline{V}$-object as $\underline{A}(A,B)$ with, possibly, a coarser uniformity. But it cannot be coarser than that induced by $\underline{A}(A,B')$, else the induced map would not be uniform. But that means that the uniformity on the pullback is the same as that of $\underline{A}(A,B)$ and so they are isomorphic.

Corollary 2. If $\{B \rightarrow B_\omega\}$ is a family such that $B \rightarrow \Pi B_\omega$ is an embedding, then

$$\begin{array}{ccc} \underline{V}(A,B) & \longrightarrow & \Pi\underline{V}(A,B_\omega) \\ \downarrow & & \downarrow \\ \underline{V}(|A|,|B|) & \longrightarrow & \Pi\underline{V}(|A|,|B_\omega|) \end{array}$$

is a pullback.

We may put these together to conclude,

Corollary 3. Under the hypotheses of Corollary 2, there is a canonical pullback,

$$\begin{array}{ccc} \underline{A}(A,B) & \longrightarrow & \Pi\underline{A}(A,B_\omega) \\ \downarrow & & \downarrow \\ [\,|A|,B] & \dashrightarrow & \Pi[\,|A|,B_\omega] \end{array} .$$

(3.13) Proposition. Let $A,B \in \underline{A}$, $C \in \underline{C}$. Then there is a canonical inclusion

$$\underline{A}(A,\underline{A}(C,B) \subset \underline{A}(C,\underline{A}(A,B)) .$$

Proof. As noted in (3.11) it is sufficient to show that there is a canonical inclusion

$$\underline{V}(A,A(C,B)) \subset \underline{V}(C,\underline{A}(A,B)) .$$

Let $\{C_\omega \to A\}$ and $\{B \to D_\psi\}$ dominate A and represent B, respectively. Then we know from 3.9 that the family $\{\underline{A}(A,B) \longrightarrow (C_\omega,D_\psi)\}$ represents $\underline{A}(A,B)$. Then from Corollary 2 above there is a canonical pullback

$$\begin{array}{ccc} \underline{V}(C,\underline{A}(A,B)) & \longrightarrow & \Pi\underline{V}(C,(C_\omega,D_\psi)) \\ \downarrow & & \downarrow \\ \underline{V}(|C|,\underline{A}(A,B)) & \longrightarrow & \Pi\underline{V}(|C|,(C_\omega,D_\psi)) \end{array}$$

and it is thus sufficient to find canonical maps $\underline{V}(A,\underline{A}(C,B)) \longrightarrow \underline{V}(C,(C_\omega,D_\psi))$ for all ω,ψ and $\underline{V}(A,\underline{A}(C,B)) \longrightarrow \underline{V}(|C|, \underline{A}(A,B))$. The fact that they are canonical implies the commutation of the square. The first is the composite

$$\underline{V}(A,\underline{A}(C,B)) \longrightarrow \underline{V}(C_\omega,(C,D_\psi)) \cong \underline{V}(C,(C_\omega,D_\psi))$$

and the second the composite

$$\underline{V}(A,\underline{A}(C,B)) \to \underline{V}(A,[|C|,B] \cong (|C|,\underline{V}(A,B))$$

where the first map comes from Corollary 4 of (3.5) and the second is the cotensor adjunction.

Corollary. A map $A \to \underline{A}(C,B)$ exchanges to a map $C \to \underline{A}(A,B)$.

(3.14) Example. We again consider the category of topological vector spaces over the discrete field K. For both possible choices of $\underline{C}$ and $\underline{D}$, the topology on a (C,D) is discrete which is that of uniform convergence on all of C. Thus $\Phi(C)$ consists of all subsets of C. When $\underline{C} = \underline{D} =$ finite dimensional spaces, it will follow from the later development that the functor $\underline{A}(-,-)$ described in this section already gives a *-autonomous structure. What we will do here is to show that when $\underline{C}$ = linearly compact and $\underline{D}$ = discrete spaces, we do not get a closed category. For A fixed the functor $\underline{V}(A,-) = |\underline{A}(A,-)|$ commutes with limits. Thus to show that $\underline{A}(A,-)$ commutes with limits, it is sufficient to show it has the right topology. But this follows from (3.9) It is now an easy application of the special adjoint functor theorem to show that $\underline{A}(A,-)$ has an adjoint which we will designate $-\otimes A$. Since $\underline{A}(A,-)$ is a $\underline{V}$-functor, the adjunction is strong which means there is a natural equivalence $\underline{V}(A\otimes B,E) = \underline{V}(A,(B,E))$ and if $\underline{A}(-,-)$ is to be a closed category structure this will have to lift to a natural equivalence

$$\underline{A}(A\otimes B,E) \cong \underline{A}(A,(B,E)) .$$

Let us now suppose that this equivalence were valid. Then for $E = K$, we get $(A\otimes B)^* \cong$

$\underline{A}(A,B^*)$. Let $A = K^T$ and $B = K^S$ with S and T infinite sets. Then $(A,B^*) \cong \underline{A}(K^T,S\cdot K) \cong (T\times S)\cdot K$ so that $(A\otimes B)^{**} = K^{S\times T}$. This implies, since every space is quasi-reflexive, that $|A\otimes B| \cong |K^{S\times T}|$. Looking ahead, we will see in the next chapter the topology on $A\otimes B$ is coarser than that of $(A\otimes B)^{**}$, i.e. that $(A\otimes B)^{**} \longrightarrow A\otimes B$ is continuous (= uniform). Since every map from a linearly compact space is closed, this is an isomorphism and so $A\otimes B \cong K^{S\times T}$. The algebraic tensor product of $|A|$ and $|B|$ consists of those functions $S\times T \longrightarrow K$ which have the form $(s,t) \longmapsto \sum_{i=1}^{n} f_i(s)g_i(t)$, $f_i \in K^S$, $g_i \in K^T$ which, when S and T are infinite, is not all of $K^{S\times T}$. Now let E be the algebraic tensor product equipped with the topology induced by $K^{S\times T}$. The identity map

$$|K^S|\otimes|K^T| \longrightarrow |E|$$

transposes to a map

$$|K^S| \longrightarrow \underline{V}(|K^T|,|E|) .$$

On the other hand, the isomorphism

$$K^S\otimes K^T \longrightarrow K^{S\times T}$$

constructed above on the hypothesis that $\underline{A}$ was a closed category transposes to a canonical map

$$K^S \longrightarrow \underline{A}(K^T, K^{S\times T})$$

which gives

$$|K^S| \longrightarrow \underline{V}(K^T, K^{S\times T}) .$$

From (3.12) we get a map

$$|K^S| \longrightarrow \underline{V}(K^T,E) .$$

Since $\underline{A}(K^T,E)$ is embedded in $\underline{A}(K^T, K^{S\times T})$ the above map $K^S \longrightarrow \underline{A}(K^T, K^{S\times T})$ factors as a map

$$K^S \longrightarrow \underline{A}(K^T, E)$$

which transposes to

$$K^S\otimes K^T \longrightarrow E .$$

Since $K^S\otimes K^T \cong K^{S\times T}$ this implies that $E \cong K^{S\times T}$ which is, as noted above, false. Thus $\underline{A}(-,-)$ does not give an internal hom.

(3.15) The counterexample suggests its own resolution. The problem arises out of the necessity of the tensor product of two objects of $\underline{C}$ to lie in $\underline{C}$. This suggests at first that we stick to the full subcategory of $\underline{A}$ consisting of complete objects. For in that case we would be forced to use a completed tensor product. Since, as is more-or-less obvious, the E constructed above is the actual tensor product, its completion is $K^{S\times T} \in \underline{C}$ as required. However, this leads to another diffficulty. We cannot be sure that the dual of a complete object is complete. It could be completed only by discontinuous (or non-uniform) maps which is obviously undesirable and would cause other difficulties as well. This problem is solved by using a modified notion of completeness which does not get in the way as much with the duality.

CHAPTER 1 CATEGORY $\underline{G}$.

1. Completeness.

(1.1) We suppose henceforth that the objects of $\underline{C}$ as well as these of $\underline{D}$ are complete in their uniformity. From this it follows that if A is embedded in ΠD_ψ, then the closure of A in that embedding is also complete and is, in fact, the uniform completion $\tilde{A}$ of A. The hypotheses of I.3.10 suffice to guarantee that $\tilde{A}$ is admissible so it belongs to $Un\underline{V}$ and then evidently to $\underline{A}$.

In addition we suppose that if $A \in \underline{A}$ is a proper closed subobject of a $C \in \underline{C}$ then the induced map $C^* \longrightarrow A^*$ (or equivalently $\underline{V}(C,T) \longrightarrow \underline{V}(A,T)$) is not injective.

(1.2) Since a completeable object is densely embedded in its completion, it is sufficient, in deciding whether every diagram

$$\begin{array}{ccc} B_1 & \longrightarrow & B_2 \\ \downarrow & & \\ A & & \end{array}$$

in which $B_1 \longrightarrow B_2$ is a dense embedding can be completed to a commutative diagram by a map $B_2 \dashrightarrow A$. To see this take $B_1 = A$, $B_2 = \tilde{A}$ and the map between them the inclusion. Take the identity for the vertical map. If there is a retraction $\tilde{A} \longrightarrow A$ which is the identity on A, the maps $\tilde{A} \longrightarrow A \rightarrow \tilde{A}$ and the identity on $\tilde{A}$ agree on A. Since A is dense in $\tilde{A}$, they are equal and the inclusion of $A \longrightarrow \tilde{A}$ is an isomorphism.

Later in these notes we will have occasion to consider weaker notions of completeness gotten by restricting the class of $B_1 \longrightarrow B_2$ for which a fill-in is required. Let us suppose that μ is a class of dense embeddings. We say that an object A is μ-complete if for every pseudomap $m : B_1 \dashrightarrow B_2$ in μ and every $f : B_1 \longrightarrow A$, there is a $g : B_2 \longrightarrow A$ (necessarily unique since the embedding is dense) such that $gm = f$. The main point of this generality follows.

(1.3) Proposition. Suppose A is μ-complete and A^Δ has a convergence uniformity (see (II.3.7)). Then A^Δ is μ-complete.

Proof. We have that $|A^\Delta| \longrightarrow |A|$ is an isomorphism in $\underline{V}$. We may thus identify the elements of A^Δ as these of A. The obvious thing to do here is to use (I.2.5). To do that we have to show that there is a basis of uniform covers of A^Δ by sets closed in A. If $\underline{d}$ is the set of pseudometrics on T, then a basic uniform cover of A^Δ is

$$\{\Gamma(a,d,S) \mid a \in A\}$$

where $S \in \Phi(A^*)$, $d \in \underline{d}$, $a \in A$ and $\Gamma(a,d,S) = \{b \mid d(\varphi a, \varphi b) < 1 \text{ for all } \varphi \in S\}$. This is refined by the uniform cover

$$\{\Gamma^\#(a,d,S) \mid a \in A\}, \text{ where}$$

$$\Gamma^\#(a,d,S) = \{b \mid d(\varphi a,\varphi b) \le \tfrac{1}{2}, \text{ for all } \varphi \in S\} = \bigcap_{\varphi \in S} \{b \mid d(\varphi a, \varphi b) \le \tfrac{1}{2}\}$$

which is an intersection of closed sets and therefore closed. To see that observe that $\{b \mid d(\varphi a, \varphi b) \le \frac{1}{2}\}$ is the inverse image of $[0, \frac{1}{2}]$ under the map

$$A \xrightarrow{\varphi} T \xrightarrow{(\varphi a, Id)} T\times T \xrightarrow{d} \mathbb{R} \quad .$$

Moreover $\Gamma^{\#}(a,d,S)$ is refined by $\Gamma(a,2d,S)$ and is thus a uniform cover.

(1.4) Given a class μ of dense embeddings and an object $A \in \underline{A}$, let μA denote the intersection of all μ - complete subobjects of $\tilde{A}$ which contain A. We wish to show that a map (respectively pseudomap) $f : A \longrightarrow B$ induces a map (respectively pseudomap) $\mu A \longrightarrow \mu B$. It does induce such a function $\tilde{A} \longrightarrow \tilde{B}$ so that we have a commutative square

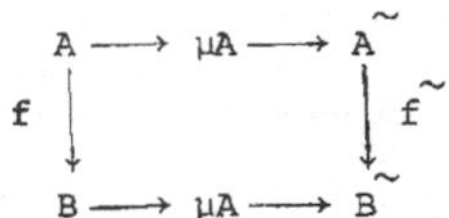

and what we need is a fill-in the middle. For this it is clearly sufficient to show that $\tilde{f}^{-1}(\mu B)$ is a μ-complete subobject of $\tilde{A}$ for it certainly contains A and hence μA. So let $E_1 \longrightarrow E_2$ belong to μ and $g : E_1 \longrightarrow \tilde{f}^{-1}(\mu B)$ be a map. First observe that g possesses a unique extension $h : E_2 \longrightarrow \tilde{A}$ since $\tilde{A}$ is complete and E_1 is densely embedded in E_2. In addition there is a pseudomap $k : E_2 \longrightarrow \mu B$ which extends $\tilde{f}g$. Restricted to the dense subset E_1 both k and $\tilde{f}k$ extend $\tilde{f}g$ and hence are equal everywhere. Since k takes values on μB, so does $\tilde{f}h$ which means the image of h lies in $\tilde{f}^{-1}(\mu B)$. Thus we have proved,

(1.5) Proposition. The object function $A \longmapsto \mu A$ extends to a $\underline{V}$-functor on $\underline{A}$.

(1.6) We let $\mu\underline{A}$ denote the full subcategory which is the image of μ. The inclusion $\mu\underline{A} \longrightarrow \underline{A}$ has a left adjoint which we find convenient to also denote by μ. The inclusion is $\underline{V}$-full and faithful and hence the adjointness is $\underline{V}$-enriched. If $A \in \mu\underline{A}$, then by the same argument used in (II.1.4), $[V,A] \in \mu\underline{A}$ for any $V \in \underline{V}$.

(1.7) We now suppose that ζ is a given class of dense embeddings. Unless there is specific mention to the contrary, ζ is taken to be the class of dense embeddings $A \longrightarrow C$ for which $C \in \underline{C}$. In one example, however (see (IV.4)ff.) this class does not satisfy hypothesis v) below. We suppose the following

i) Every dense embedding $A \longrightarrow C$ with $C \in \underline{C}$ belongs to ζ;

ii) If $A \longrightarrow B$ belongs to ζ, then B is embedded in an object of $\underline{C}$;

iii) If $A \longrightarrow B$ belongs to ζ, then B^* is complete;

iv) If $A \longrightarrow B$ belongs to ζ, then B is prereflexive;

v) Every ζ-complete object is quasi-reflexive.

We note that should $\underline{C}$ be stable under the formation of closed subobjects - it usually is - then ζ-completeness for any class ζ satisfying the above is equivalent to $\zeta^{\#}$-completeness for the class $\zeta^{\#}$ of dense embeddings $A \longrightarrow C$ with $C \in \underline{C}$. For if $A \to B \in \zeta$ and B is embedded in $C \in \underline{C}$, then the closure of A in C is its uniform completion $\tilde{A}$. The possibility of extending a map defined on A to all of $\tilde{A}$ certainly implies the possibility of extending it to B.

(1.8) We say that A is ζ-$*$-complete provided A^* is ζ-complete.

(1.9) Proposition. Let A be ζ-complete. Then $\delta A = (\zeta A^*)^*$ is ζ-complete, ζ-$*$-complete and reflexive. Moreover there is a canonical bijection $\delta A \longrightarrow A$, meaning that δA has the same underlying $\underline{V}$-object with a possibly finer uniformity.

Proof. The inclusion $A^* \longrightarrow \zeta A^*$ induces $\delta A = (\zeta A^*)^* \longrightarrow A^{**}$. Since $A^* \longrightarrow \zeta A^*$ is a dense inclusion and $T \in \underline{D}$ is complete, the induced map $\underline{V}(\zeta A^*, T) \longrightarrow \underline{V}(A^*, T)$ is an isomorphism. That is, $|\delta A| \xrightarrow{\cong} |A^{**}|$. Since A is ζ-complete, it is by hypothesis quasi-reflexive which means that $A^{**} = A^{\Delta}$ and so $A^{**} \longrightarrow A$ exists and is bijective. Thus there is a canonical bijection $\delta A \longrightarrow A$. Now let $B_1 \longrightarrow B_2$ belong to ζ and $B_1 \rightarrow \delta A$ be given. We get, by composition

$$B_1 \longrightarrow A$$

$$B_2 \longrightarrow A$$

since A is ζ-complete,

$$A^* \longrightarrow B_2^*$$

$$\zeta A^* \longrightarrow B_2^*$$

since B_2^* is assumed complete,

$$B_2^{**} \longrightarrow \delta A$$

$$B_2 \longrightarrow \delta A$$

since B_2 is assumed prereflexive meaning there is a canonical map $B_2 \longrightarrow B_2^{**}$. It is easy to check that the restriction to B_1 is the given map $B_1 \longrightarrow \delta A$. This shows that δA^* is ζ-complete. Since ζA^* is ζ-complete, it is evidently quasi-reflexive. Thus $(\delta A)^* = (\zeta A^*)^{**} = (\zeta A^*)^{\Delta} \longrightarrow \zeta A^*$ which upon dualizing gives the canonical map

$$\delta A \longrightarrow (\delta A)^{**}$$

and shows that δA is pre-reflexive. As it also is ζ-complete, hence quasi-reflexive it is reflexive. Finally, ζA^* is ζ-complete, hence so is $(\zeta A^*)^{\Delta} = (\zeta A^*)^{**} = (\delta A)^*$ so that δA is ζ-$*$-complete.

(1.10) Proposition. Let μ be a class of dense inclusions and B be μ-complete. Then for any $C \in \underline{C}$, $\underline{A}(C,B)$ is μ-complete.

Proof. Let $\{B \rightarrow D_\omega\}$ be a $\underline{D}$-representation of B. Then $\{\underline{A}(C,B) \longrightarrow (C,D_\omega)\}$ is a $\underline{D}$-representation of $\underline{A}(C,B)$ (see II.3.10, Corollary). For any $E \in \underline{A}$ the diagram

$$\begin{array}{ccc} \underline{V}(E,\underline{A}(C,B)) & \longrightarrow & \Pi \underline{V}(E,(C,D_\omega)) \\ \downarrow & & \downarrow \\ \underline{V}(E,[\,|C|\,,B]\,) & \longrightarrow & \Pi \underline{V}(E,[\,|C|\,,D_\omega]) \end{array}$$

is a pullback (see II.3.12, Corollary 1 and apply $\underline{V}(E,-)$). If $E_1 \longrightarrow E_2$ belongs to μ, it induces isomorphisms $\underline{V}(E_2,(C,D_\omega)) \longrightarrow \underline{V}(E_1,(C,D_\omega))$,

$$\underline{V}(E_2,[\,|C|\,,B\,]) \cong \underline{V}(\,|C|\,,\underline{V}(E_2,B))$$

$$\cong \underline{V}(\,|C|,\underline{V}(E_1,B)) \cong \underline{V}(E_1,[\,|C|,B]\,),$$

$$\underline{V}(E_2,[\,|C|,D_\omega]) \cong \underline{V}(|C|,\underline{V}(E_2,D_\omega))$$

$$\cong \underline{V}(|C|,\underline{V}(E_1,D_\omega)) \cong \underline{V}(E_1,[\,|C|,D_\omega]) \quad ,$$

since (C,D_ω), B and D_ω are all μ - complete. But then three of the four vertices are the same whether $E = E_1$ or $E = E_2$ and hence so is the fourth (the pullback).

Corollary. Let A be reflexive and ζ-*-complete. Then for any $D \in \underline{D}$, $\underline{A}(A,D)$ is ζ-complete.

Proof. Dualize and use (II.3.6).

(1.11) Theorem. Let A be reflexive and ζ-*-complete and B ζ-complete. Then $\underline{A}(A,B)$ is ζ-complete.

Proof. Proceed exactly as in the proof of (1.10) except replace C by A . The pullback is

$$\begin{array}{ccc} \underline{V}(E,\underline{A}(A,B)) & \longrightarrow & \Pi\underline{V}(E,\underline{A}(A,D_\omega)) \\ \downarrow & & \downarrow \\ \underline{V}(E,[\,|A|,B]) & \rightarrowtail & \Pi\underline{V}(E,[\,|A|,D_\omega]) \end{array}$$

with the above corollary providing the necessary ζ-completeness of $\underline{A}(A,D_\omega)$.

(1.12) Example. We return once more to the example of vector spaces. When $\underline{C}$ and $\underline{D}$ are the finite dimensional spaces, topologized discretely, then the spaces are certainly complete. The definition of ζ-completeness requires that we begin with a subspace of a space in $\underline{C}$ - necessarily finite dimensional - and a dense subspace - necessarily the whole thing. Thus the required map extension property is trivially satisfied and every space is ζ-complete. As well, then, is every space ζ-*-complete. The situation is quite different when $\underline{D}$ consists of discrete, and $\underline{C}$ of linearly compact spaces. Those of $\underline{D}$ are evidently complete as are those of $\underline{C}$ which are products of discrete spaces. Not every space is ζ-complete. Any dense proper subspace of a linearly compact space cannot be ζ-complete. For example the subspace of K^S, S infinite, consisting of the elements of the S-fold direct sum is not ζ-complete. If we call that space V , we have

$$S\cdot K \longrightarrow V \longrightarrow K^S$$

with the first map bijective and the second a dense embedding. Dualization gives

$$S\cdot K \longrightarrow V^* \longrightarrow K^S \quad .$$

The first map being the dual of a dense embedding is a bijection and the second is evidently dense. We know it is dense as soon as we know that V does not contain any infinite dimensional linear compact subspaces. For then all the $C \to V$ are with C discrete and factor through $S\cdot K$, whence $(S\cdot K)^*$ and V^* are embedded in the sum space. This is presumably always true but certainly is if S and K are both countably infinite. For any infinite dimensional linearly compact space, being a power of K , is uncountable while $S\cdot K$ is countable. Thus in that case $V^* \cong V$ and so $V^{**} \cong V$. The former isomorphism is induced by the standard inner product and hence the latter one is induced by the canonical map. This shows that V is not ζ-$*$-complete either, but that V is reflexive. If V is closed in W, W/V is a separated linearly topologized space and hence has continuous functionals. Thus there are non-zero functionals on W which vanish on V, confirming the hypothesis made in (1.1).

2. Definition and Elementary Properties of $\underline{G}$.

(2.1) It will be a standing hypothesis in this chapter that ζ-complete objects are quasireflexive. The demonstrations of this fact in the various examples seem quite different - save for the cases in which (II.2.8) is satisfied - and do not seem to have any common generalization. It is clear that the hypotheses of (II.2.8) are not always satisfied.

(2.2) We let $\underline{G}$ denote the full subcategory of $\underline{A}$ consisting of objects which are reflexive, ζ-complete and ζ-$*$-complete. It follows from hypotheses we have made that both $\underline{C}$ and $\underline{D}$ are subcategories of $\underline{G}$. Also $\underline{G}$ is complete and cocomplete. In fact $\mathrm{Un}\underline{V}$ is complete, essentially by hypothesis (see (I.3.10)) and cocomplete by the adjoint functor theorem. $\underline{A}$ being reflexive is also complete and cocomplete as is $\zeta\underline{A}$. The last step follows from the next proposition.

(2.3) Proposition. The inclusion $\underline{G} \longrightarrow \zeta\underline{A}$ has a right adjoint $A \mapsto \delta A$.
Proof. That δA belongs to $\underline{G}$ follows from (1.9). Now if $G \in \underline{G}$ and $G \to A$ is a map, we have $A^* \longrightarrow G^*$ and with G^* ζ-complete, we have $\zeta A^* \longrightarrow G^*$ whence $G \cong G^{**} \longrightarrow (\zeta A^*)^*$. Uniqueness follows from the fact that the maps $(\zeta A^*)^* \longrightarrow A^{**} \longrightarrow A$ are bijective, the first because A^* is densely embedded in ζA^* and the second because A is ζ-complete, hence quasi-reflexive.

(2.4) Let $C_1, C_2 \in \underline{C}$. The identity map

$$(C_1, C_2^*) \longrightarrow (C_1, C_2^*)$$

transposes, by the corollary of II.3.13 to a map

$$C_1 \longrightarrow ((C_1, C_2^*), C_2^*) \cong (C_2, (C_1, C_2^*)^*) .$$

Applying $|-|$ we get a map in $\underline{V}$

$$|C_1| \longrightarrow \underline{V}(C_2, (C_1, C_2^*)^*) \longrightarrow \underline{V}(|C_2|, |(C_1, C_2^*)^*|)$$

$$|C_1| \otimes |C_2| \longrightarrow |(C_1, C_2^*)^*| .$$

Let $\tau(C_1, C_2)$ denote the image of that map as an embedded subobject of $(C_1, C_2^*)^*$.

(2.5) Proposition. For any $C_1, C_2 \in \underline{C}$, $\tau(C_1, C_2)$ is dense in $(C_1, C_2^*)^*$.
Proof. (C_1, C_2^*) lies in $\underline{D}$ so its dual is in $\underline{C}$. By the hypothesis made in (1.1) if $\tau(C_1, C_2)$ is not dense, its closure is proper and the induced map

$$\underline{V}(C_1, C_2^*) \longrightarrow \underline{V}(\tau(C_1, C_2), T)$$

is not injective. But the map

$$\underline{V}(C_1, C_2^*) \longrightarrow \underline{V}(\tau(C_1, C_2), T) \longrightarrow \underline{V}(|C_1| \otimes |C_2|, |T|)$$

can also be factored as

$$\begin{aligned} &\underline{V}(C_1,C_2^*) \dashrightarrow \underline{V}(|C_1|,|C_2^*|) \\ \cong\ &\underline{V}(|C_1|,\underline{V}(C_2,T)) \longrightarrow \underline{V}(|C_1|,\underline{V}(|C_2|,|T|)) \\ \cong\ &\underline{V}(|C_1|\otimes|C_2|,|T|)\ , \end{aligned}$$

each term of which is injective. Since the first factor of an injection is an injection, the result follows.

(2.6) Proposition. Let C_1, $C_2 \epsilon \underline{C}$. For any $A \epsilon \underline{A}$ there is a canonical map

$$\underline{V}(C_1,\underline{A}(C_2,A)) \dashrightarrow \underline{V}(|\tau(C_1,C_2)|,|A|)$$

Proof. Let $\{A \longrightarrow D_\omega\}$ be a $\underline{D}$-representation of A. Since $|\tau(C_1,C_2)|$ is a regular image of $|C_1|\otimes|C_2|$, there is a pullback

$$\begin{array}{ccc} \underline{V}(|\tau(C_1,C_2)|,|A|) & \longrightarrow & \Pi\underline{V}(|\tau(C_1,C_2)|,|D_\omega|) \\ \downarrow & & \downarrow \\ \underline{V}(|C_1|\otimes|C_2|,|A|) & \dashrightarrow & \Pi\underline{V}(|C_1|\otimes|C_2|,|D_\omega|) \end{array} .$$

Then map

$$\begin{aligned} &\underline{V}(C_1,\underline{A}(C_2,A)) \longrightarrow \Pi\underline{V}(C_1,(C_2,D_\omega)) \\ \longrightarrow\ &\Pi\underline{V}(C_1,(D_\omega^*,C_2^*)) \longrightarrow \Pi\underline{V}(D_\omega^*,(C_1,C_2^*)) \\ \longrightarrow\ &\Pi\underline{V}((C_1,C_2^*)^*,D_\omega) \longrightarrow \Pi\underline{V}(|(C_1,C_2^*)^*|,|D_\omega|) \\ \longrightarrow\ &\Pi\underline{V}(|\tau(C_1,C_2)|,|D_\omega|) \quad . \end{aligned}$$

Also map

$$\begin{aligned} &\underline{V}(C_1,\underline{A}(C_2,A)) \longrightarrow \underline{V}(|C_1|,\underline{V}(C_2,A)) \\ \longrightarrow\ &\underline{V}(|C_1|,\underline{V}(|C_2|,|A|)) \cong \underline{V}(|C_1|\otimes|C_2|,|A|) \quad . \end{aligned}$$

Since each of these maps is canonical, they give the same map to $\Pi\underline{V}(|C_1|\otimes|C_2|,|D_\omega|)$ and hence the map required.

(2.7) Proposition. Let C_1, $C_2 \epsilon \underline{C}$, $A \epsilon \underline{A}$. Then there is a canonical map

$$\underline{V}(C_1,\underline{A}(C_2,A)) \longrightarrow \underline{V}(\tau(C_1,C_2),A) \quad .$$

Proof. Let $\{A \longrightarrow D_\omega\}$ be a $\underline{D}$-representation of A. By (II. 3.12), there is a pullback

$$\begin{array}{ccc} \underline{V}(\tau(C_1,C_2),A) & \longrightarrow & \Pi\underline{V}(\tau(C_1,C_2),D_\omega) \\ \downarrow & & \downarrow \\ \underline{V}(|\tau(C_1,C_2)|,|A|) & \longrightarrow & \Pi\underline{V}(|\tau(C_1,C_2)|,|D_\omega|) \end{array} .$$

We map

$$\begin{aligned} &\underline{V}(C_1,\underline{A}(C_2,A)) \longrightarrow \Pi\underline{V}(C_1,(C_2,D_\omega)) \\ \cong\ &\Pi\underline{V}(C_1,(D_\omega^*,C_2^*)) \cong \Pi\underline{V}(D_\omega^*,(C_1,C_2^*)) \\ \cong\ &\Pi\underline{V}((C_1,C_2^*)^*,D_\omega) \longrightarrow \Pi\underline{V}(\tau(C_1,C_2),D_\omega) \end{aligned}$$

while (2.6) provides $\underline{V}(C_1,\underline{A}(C_2,A)) \longrightarrow \underline{V}(|\tau(C_1,C_2)|,|A|)$ from which we have the required map.

Corollary. If, in addition, A is ζ-complete, there is a canonical map

$$\underline{V}(C_1,\underline{A}(C_2,A)) \longrightarrow \underline{V}((C_1,C_2^*)^*,A)) \quad .$$

(2.8) Proposition. Let $C_1,C_2 \in \zeta\underline{A}$. Then the canonical map

$$\underline{V}(C_1,\underline{A}(C_2,\delta A)) \longrightarrow \underline{V}(C_1,\underline{A}(C_2,A))$$

is an isomorphism.

Proof. The inverse of the canonical map is the composite

$$\underline{V}(C_1,\underline{A}(C_2,A)) \longrightarrow \underline{V}((C_1,C_2^*)^*,A)$$
$$\longrightarrow \underline{V}(A^*,(C_1,C_2^*)) \cong \underline{V}(\zeta A^*,(C_1,C_2^*))$$
$$\longrightarrow \underline{V}(C_1,\underline{A}(\zeta A^*,C_2^*)) \rightarrow \underline{V}(C_1,\underline{A}(C_2,\delta A))$$

where the next to last map is the $\underline{V}$-morphism underlying the map of II.3.13.

(2.9) Proposition. Let $C\in\underline{C}$, $G\in\underline{G}$, $A\in\zeta\underline{A}$. Then the canonical map

$$\underline{V}(C,\underline{A}(G,\delta A)) \longrightarrow \underline{V}(C,\underline{A}(G,A))$$

is an isomorphism.

Proof. Let $\{C_\omega \longrightarrow G\}$ be a $\underline{C}$-representation of G . Then there is a pullback

$$\begin{array}{ccc} \underline{V}(C,\underline{A}(G,\delta A)) & \longrightarrow & \Pi\underline{V}(C,\underline{A}(C_\omega,\delta A)) \\ \downarrow & & \downarrow \\ \underline{V}(|C|,\underline{V}(G,\delta A)) & \longrightarrow & \Pi\underline{V}(|C|,\underline{V}(C_\omega,\delta A)) \end{array} \quad .$$

We map

$$\underline{V}(C,\underline{A}(G,A)) \longrightarrow \Pi\underline{V}(C,\underline{A}(C_\omega,A)) \longrightarrow \Pi\underline{V}(C,\underline{A}(C_\omega,\delta A))$$

as established in (2.8). Also we have

$$\underline{V}(C,\underline{A}(G,A)) \longrightarrow \underline{V}(|C|,\underline{V}(G,A)) \cong \underline{V}(|C|,\underline{V}(G,\delta A)) \quad .$$

This gives the required map.

(2.10) Proposition. Let $C\in\underline{C}$, $D\in\underline{D}$ and $G\in\underline{G}$. Then the canonical map of (II.3.13)

$$\underline{A}(G,(C,D)) \longrightarrow \underline{A}(C,\underline{A}(G,D))$$

is an isomorphism.

Proof. As observed in (II.3.11) it is sufficient to show that

$$\underline{V}(G,(C,D)) \longrightarrow \underline{V}(C,\underline{A}(G,D))$$

is an isomorphism. The inverse of the canonical map is given by

$$\underline{V}(C,\underline{A}(G,D)) \longrightarrow \underline{V}(C,\underline{A}(D^*,G^*)) \longrightarrow \underline{V}((C,D)^*,G^*) \cong \underline{V}(G,(C,D)) \quad ,$$

the second map coming from the corollary to (2.7).

(2.11) Proposition. Let $A,B\in\underline{A}$, $G\in\underline{G}$. Then there is a canonical map

$$\underline{A}(A,\underline{A}(G,B)) \longrightarrow \underline{A}(G,\underline{A}(A,B))$$

Proof. As observed in (II.3.11) it is sufficient to have a canonical map

$$\underline{V}(A,\underline{A}(G,B)) \longrightarrow \underline{V}(G,\underline{A}(A,B)) \quad .$$

Let $\{C_\omega \longrightarrow A\}$ and $\{B \longrightarrow D_\psi\}$ be a $\underline{C}$-domination and $\underline{D}$-representation, respectively. Then there is a pullback

$$\begin{array}{ccc} \underline{V}(G,\underline{A}(A,B)) & \longrightarrow & \Pi\underline{V}(G,(C_\omega,D_\psi)) \\ \downarrow & & \downarrow \\ \underline{V}(|G|,\underline{V}(A,B)) & \longrightarrow & \Pi\underline{V}(|G|,\underline{V}(C_\omega,D_\psi)) \end{array} \quad .$$

We have a canonical map

$$\underline{V}(A,\underline{A}(G,B)) \longrightarrow \Pi\underline{V}(C_\omega,\underline{A}(G,D_\psi)) \cong \Pi\underline{V}(G,(C_\omega,D_\psi))$$

by (2.10). As well we have

$$\underline{V}(A,\underline{A}(G,B)) \longrightarrow \underline{V}(A,[|G|,B]) \cong \underline{V}(|G|,\underline{V}(A,B)) \quad ,$$

induced by the embedding $\underline{A}(G,B) \longrightarrow [\,|G|,B]$.

Corollary 1. Let $A\in\underline{A}$, G, $H\in\underline{G}$. Then there is a canonical isomorphism

$$\underline{A}(G,\underline{A}(H,A)) \cong \underline{A}(H,\underline{A}(G,A)) \quad .$$

Corollary 2. Let $A\in\zeta A$, $C\in\underline{C}$, $G\in\underline{G}$. Then the canonical map

$$\underline{V}(G,\underline{A}(C,\delta A)) \longrightarrow \underline{V}(G,\underline{A}(C,A))$$

is an isomorphism.

(2.12) Proposition. Let $A\in\zeta A$, G, $H\in\underline{G}$. Then the natural map

$$\underline{V}(H,\underline{A}(G,\delta A)) \longrightarrow \underline{V}(H,\underline{A}(G,A))$$

is an isomorphism.

Proof. Replace C by H everywhere in the proof of (2.9). The necessary isomorphism

$$\underline{V}(H,\underline{A}(C_\omega,\delta A)) \cong \underline{V}(H,\underline{A}(C_\omega,A))$$

is established in corollary 2 above.

Corollary. A map $H \longrightarrow \underline{A}(G,\delta A)$ is equivalent to a map $H \longrightarrow \underline{A}(G,A)$.

Proof. Apply $\mathrm{Hom}(I,-)$ to the above.

(2.13) Proposition. Let $A,B,E\in\zeta A$. Then there is a canonical map

$$\underline{V}(A,\underline{A}(B,E)) \longrightarrow \underline{V}(\delta A,\delta\underline{A}(\delta B,\delta E)) \quad .$$

Proof. We have

$$\begin{aligned} & \underline{V}(A,\underline{A}(B,E)) \longrightarrow \underline{V}(A,\underline{A}(\delta B,E)) \\ \longrightarrow\ & \underline{V}(\delta B,\underline{A}(A,E)) \longrightarrow \underline{V}(\delta B,\underline{A}(\delta A,E)) \\ \longrightarrow\ & \underline{V}(\delta B,\underline{A}(\delta A,\delta E)) \longrightarrow \underline{V}(\delta A,\underline{A}(\delta B,\delta E)) \\ \cong\ & \underline{V}(\delta A,\delta\underline{A}(\delta B,\delta E)) \quad . \end{aligned}$$

Here the first and third arrow are induced by back adjunctions, the second and fourth from 2.11, the fifth from 2.12 and the last isomorphism is the one given by the adjunction.

Corollary. Under the same hypotheses a map $A \longrightarrow \underline{A}(B,E)$ gives a map $\delta A \longrightarrow \delta A(\delta B,\delta E)$.

Proof. Apply Hom(I,-) to the above.

3. The Closed Monoidal Structure on $\underline{G}$

(3.1) Let $G,H \in \underline{G}$. Since by (1.11), $\underline{A}(G,H)$ is ζ-complete, $\delta\underline{A}(G,H) \in \underline{G}$. We define $\underline{G}(G,H) = \delta\underline{A}(G,H)$. Note that since $\delta A \longrightarrow A$ is bijective for all $A \in \zeta\underline{A}$, it follows that $|\underline{G}(G,H)| \cong \underline{V}(G,H)$. In particular every element of $\underline{G}(G,H)$ is a uniform pseudomap. This explains the principal advantage of using ζ-completeness instead of completeness. No non-uniform maps need be added to hom object to make it ζ-complete.

(3.2) Proposition. Let $G,H \in \underline{G}$. Then there is an evaluation map $G \longrightarrow \underline{A}(\underline{A}(G,H),H)$.

Proof. By (2.11) there is a canonical map

$$\underline{A}(\underline{A}(G,H),\underline{A}(G,H)) \longrightarrow \underline{A}(G,\underline{A}(\underline{A}(G,H),H)) .$$

The required map is, of course, the image of the identity.

(3.3) Proposition. Let $G,H,K \in \underline{G}$. Then composition of morphisms determines a canonical map

$$\underline{A}(G,H) \longrightarrow \underline{A}(\underline{A}(H,K),\underline{A}(G,K)) .$$

Proof. From the discussion following (II.3.11) it follows that there is a single $A \in \underline{A}$ into which both $\underline{A}(G,\underline{A}(\underline{A}(H,K),K))$ and $\underline{A}(\underline{A}(H,K),\underline{A}(G,K))$ may be embedded. Moreover, from (II.3. 12) it follows that

$$\begin{array}{ccc} \underline{V}(\underline{A}(G,H),\underline{A}(\underline{A}(H,K),\underline{A}(G,K))) & \longrightarrow & \underline{V}(\underline{A}(G,H),A) \\ \downarrow & & \downarrow \\ \underline{V}(\underline{V}(G,H),\underline{V}(\underline{A}(H,K),\underline{A}(G,K))) & \longrightarrow & \underline{V}(\underline{V}(G,H),|A|) \end{array}$$

is a pullback. Now the preceding proposition provides a map

$$H \longrightarrow \underline{A}(\underline{A}(H,K),K)$$

to which we may apply the functor $\underline{A}(G,-)$ to get

$$\underline{A}(G,H) \longrightarrow \underline{A}(G,\underline{A}(\underline{A}(H,K),K)) \longrightarrow A ,$$

which is an element of $\underline{V}(\underline{A}(G,H),A)$. From (II.3.4), we have a map

$$\underline{V}(G,H) \longrightarrow \underline{V}(\underline{A}(H,K),\underline{A}(G,K))$$

which expresses the fact that $\underline{A}(-,K)$ is a $\underline{V}$-functor. This is a canonical element of

$$\underline{V}(\underline{V}(G,H),\underline{V}(\underline{A}(G,H),\underline{A}(G,K)))$$

and both being canonical give the same element of $\underline{V}(\underline{V}(G,H),|A|)$. Thus we get an element of $\underline{V}(\underline{A}(G,H),\underline{A}(\underline{A}(G,H),\underline{A}(G,K)))$. Since the elements we began with were maps and Hom(I,-) commutes with pullbacks, this element too is a map

$$\underline{A}(G,H) \longrightarrow \underline{A}(\underline{A}(G,H),\underline{A}(G,K)) \quad .$$

(3.4) Proposition. Let $G,H,K \in \underline{G}$. Then composition gives maps

$$\underline{G}(G,H) \longrightarrow \underline{G}(\underline{G}(H,K),\underline{G}(G,K))$$

$$\underline{G}(H,K) \longrightarrow \underline{G}(\underline{G}(G,H),\underline{G}(G,K)) \quad .$$

Proof. The first comes by applying the corollary of (2.11) to the above. The second comes from transposing $\underline{G}(G,H) \longrightarrow \underline{G}(\underline{G}(H,K),\underline{G}(G,K)) \longrightarrow (\underline{G}(H,K),\underline{G}(G,K))$ to get $\underline{G}(H,K) \longrightarrow \underline{A}(\underline{G}(G,H),\underline{G}(G,K))$ and applying ζ once more.

(3.5) Theorem $\underline{G}$ is $\star$-autonomous.

Proof. Since $I \in \underline{C}$, $I \in \underline{G}$. If $G \in \underline{G}$, $G^* \cong (G,I) \in \underline{G}$ and hence $G^* = \underline{G}(G,I)$. If $G \longrightarrow \underline{G}(H,K^*)$, then we have $G \longrightarrow \underline{G}(H,K^*) \longrightarrow \underline{G}(K,H^*)$ and $K \longrightarrow \underline{G}(G,H^*) \cong \underline{G}(H,G^*)$. Then by I.4.4 we have all the data required.

4. Summary of the Hypotheses.

(4.1) The hypotheses used in this construction are rather complicated. In addition they are scattered all over the preceding three chapters, being introduced as needed. Thus it seems useful to collect in one place a summary of these hypotheses and a reference to a fuller exposition of each. It is understood that these are the assumptions under which (3.4) is proved. In some cases more restrictive forms of these hypotheses are stated that are satisfied in some of the examples. This will be mentioned in the examples.

(4.2) The hypotheses are

(i) $\underline{V}$ is an autonomous (i.e. closed, symmetric, monoidal) category (see (I.1.1), (I.1.2)) .

(ii) $\underline{V}$ is a semi-variety; that is a full subcategory of a variety closed under projective limits and containing all the free algebras (see (I.1.4) - (I.1.8)) and that the hypothesis of (I.3.10) is satisfied.

(iii) The subcategories $\underline{C}$ and $\underline{D}$ of Un $\underline{V}$ (see (I.3)) have the structure of a pre-$\star$-autonomous situation (see I.4.6.) and that every $C \in \underline{C}$ has a $\underline{D}$-representation (see (II.1.2)); moreover the uniformity on the objects (C,D) is a convergence uniformity (see (II.3.7)) .

(iv) Every ζ-complete object (see (III.1.4) and (III.1.7)) is prereflexive (see (II.2.7) also (II.2.9)).

(v) For every $C \in \underline{C}$ and every closed proper subobject $A \longrightarrow C$, the map $C^* \longrightarrow A^*$ is not injective; equivalently $C^* \longrightarrow A^*$ injective iff A is dense in C (see(III.1.1)).

(vi) Every object in $\underline{C}$ and every object of $\underline{D}$ is complete.

Before getting into examples we need to know that for any ring R, the category Top Mod R of topological R-modules is equivalent to Un Mod R of uniform R-modules.

In fact, let A be a topological module. If M is a neighborhood of 0 in A, let $\underline{u}(M) = \{a+M \mid a\in A\}$. The collection of all $\underline{u}(M)$ as M ranges over all neighborhoods of 0, is a uniformity. For if $N-N \subset M$, I claim $\underline{u}(N)$ is a star refinement of $\underline{u}(M)$. To see that, suppose $b\in \underline{u}^*(a,N)$. Then

$$b\in\bigcup\{c+N \mid a\in c+N\}$$

which means there is a $c\in A$ with $a\in c+N$, $b\in c+N$ so that $b-a\in N-N \subset M$ or $b\in a+M$. Next I claim that subtraction is uniform. In fact, if $N-N \subset M$, the image under subtraction of $\underline{u}(N)\times\underline{u}(N)$ refines $\underline{u}(M)$. That scalar multiplication is uniform is a special case of the fact that a continuous map between topological modules induces a uniform one. In fact if $f : A' \longrightarrow A$ is continuous, and M a neighborhood of 0 in A then $f^{-1}(\underline{u}(M)) = \underline{u}(f^{-1}(M))$.

To go the other way, just use the standard functor from uniform spaces to topological spaces described in I.2. Since that functor preserves products ([Isbell], I.8, p.17), the uniformity of the operations implies their continuity. It is clear that the uniform topology of a uniform group constructed from a topological group is the original topology. To go the other way, we must show that every uniform cover in a uniform group is refined by cover by translates of a neighborhood of 0. So let G be a uniform group and let $\underline{u}$ be a uniform cover. By the uniformity of addition there is a uniform cover $\underline{v}$ such that $V_1, V_2 \in \underline{v}$ implies there is a $U\in\underline{u}$ with $V_1+V_2 \subset U$. There is a neighborhood M of 0 in $\underline{v}$. (Let M contain $st(0,\underline{w})$ for some star refinement of $\underline{v}$.) Then for all $V\in\underline{v}$, V+M is in some set in $\underline{u}$. In particular for $a\in V$, a+M is in some set in $\underline{u}$ and so

$$\{a+M \mid a\in A\}$$

refines $\underline{u}$.

1. Vector Spaces.

(1.1) The theory for topological vector spaces over a discrete field has been explicated completely in the earlier Chapters. However, for special choices of K, it is clear that other choices for $\underline{C}$ and $\underline{D}$ will result in different theories. In what follows we take $K = \mathbb{R}$ or $K = \mathbb{C}$, always with the usual topology.

(1.2) Let $\underline{C}$ be the subcategory of Top $\underline{V}$ consisting of finite and countable powers K^S of K and $\underline{D}$ be the category of finite and countable direct sums $S\cdot K$ topologized with locally convex sum topology (explained below). The duality $\underline{C}^{op} \longrightarrow \underline{D}$ as well as its inverse $\underline{D}^{op} \longrightarrow \underline{C}$ is just the hom into K, topologized by uniform convergence on compact sets.

(1.3) In fact a compact set in K^S has compact projection on every coordinate. Thus for $s\in S$ the projection on the s coordinate is contained in the closed disc $\Delta(r_s)$

of radius r_s . Hence every compact set is a subset of $\prod_{s\in S}\Delta(r_s)$. It is evident that a continuous linear map $K^S \longrightarrow K$ is represented by an element (x_s) of $S\cdot K$ by the formula

$$(x_s)(y_s) = \Sigma x_s y_s .$$

Then the neighborhood

$$\{(x_s) \mid (x_s)\Pi\Delta(r_s) < 1\} = \{(x_s) \mid \Sigma|x_s r_s| < 1\}$$

can be easily described as follows. Let $\overset{\circ}{\Delta}(1/r_s) \subset K$ denote the open disc of radius $1/r_s$ (or all of K where $r_s = 0$) and $\Gamma\{\overset{\circ}{\Delta}(1/r_s) \mid s\in S\}$ the subset of $S\cdot K$ consisting of all elements $(\lambda_s x_s)$ such that $x_s \in \overset{\circ}{\Delta}(1/r_s)$ and $\Sigma|\lambda_s| = 1$. In other words $\Gamma\overset{\circ}{\Delta}(1/r_s)$ is the convex circled hull of the images of the $\overset{\circ}{\Delta}(1/r_s)$ in the sum. Then if $|y_s| < r_s$ we have $\Sigma|\lambda_s x_s y_s| < \Sigma|\lambda_s| = 1$. On the other hand suppose $(z_s)\in S\cdot K$ is a sequence for which $(y_s)\in\Pi\Delta(r_s)$ implies $|\Sigma y_s z_s| < 1$. Then choose for each $s\in S$ an element $y_s \in K$ of absolute value r_s such that $y_s z_s$ is real and positive. Then $\Sigma y_s z_s = r < 1$. Let $\lambda_s = r_s z_s/r$. We have $\Sigma|\lambda_s| = \Sigma|r_s z_s|/r = \Sigma y_s z_s/r = 1$. If now $x_s = z_s/x_s$, we have $|x_s| = r|z_s|/r_s|z_s| = r/r_s < 1/r_s$. This shows that $(K^S)^* \cong S\cdot K$.

(1.3) To go the other way, we first observe that algebraically $(S\cdot K)^* \cong K^S$. The only question is the topology. The product topology on K^S is that of pointwise convergence. We begin with,

<u>Proposition</u>. Every compact set in $S\cdot K$ is in a finite dimensional subspace.

Proof. Suppose Z is a compact subset of $S\cdot K$ and suppose for countably many s, say $s = 1,2,3,\ldots,n,\ldots$ there is a point $(x_s^{(n)}) \in Z$ with $x_n \neq 0$. Let $p = (p_s)$ be the seminorm on $S\cdot K$ such that $p_n(x) = n|x/x_n^{(n)}|$. Then since p_n is a continuous seminorm on K , $p = (p_n)$ is a continuous seminorm on $S\cdot K$ (this is standard result on topological vector spaces) and $p(x_s^{(n)}) \geq n$. Thus p is unbounded on Z which contradicts its compactness.

(1.4) Now a compact set in $S\cdot K$ is a compact set in some finite dimensional subspace. In fact it is contained in a space $S_0\cdot K$ for some finite subset $S_0 \subset S$. Just take S_0 as the set of all elements of S necessary to express a basis for this finite dimensional subspace. A compact set in $S_0\cdot K$ is contained in a set of the form

$$\Delta_\lambda = \{\Sigma\{\lambda_s s \mid s\in S_0\} \mid \Sigma|\lambda_s| \leq \lambda\}$$

where λ is a fixed positive real number. Now if φ is a linear functional and $\varphi(s) < \epsilon/\lambda$, $\Sigma|\lambda_s| < \lambda$ implies

$$\begin{aligned} |\varphi(\Sigma\lambda_s s)| &= |\Sigma\lambda_s \varphi(s)| \\ &< \Sigma|\lambda_s|\epsilon/\lambda \\ &\leq \epsilon \end{aligned}$$

so that the basic neighborhood of 0

$$\{\varphi \mid \varphi(\Delta_\lambda) < \lambda\} \supset \{\varphi \mid \varphi(s) < \epsilon/\lambda , s\in S_0\} ,$$

which shows that $(S\cdot K)^*$ has the topology of pointwise convergence and hence is K^S .

(1.5) The hypotheses of III.4.2, parts (i) and (ii) do not depend on the choice of $\underline{C}$ and $\underline{D}$ and thus are still satisfied. For the third part, we let, for $C \in \underline{C}$, $D \in \underline{D}$, (C,D) denote the space of continuous maps $C \longrightarrow D$ topologized by uniform convergence on compact subsets of C.

<u>Proposition</u>. Let $C = K^S$ and $D = P \cdot K$. Then

$$(C,D) \cong (S \times P) \cdot K .$$

Proof. Let $\| - \| : P \cdot K \longrightarrow \mathbb{R}$ be the norm defined by

$$\| (x_p) \| = \Sigma \, |x_p| .$$

which is defined since there are only finitely many non-zero x_p. Let $f : K^S \longrightarrow P \cdot K$ be a continuous linear map. For $s \in S$, let f_s denote the restriction of f to the s coordinate (all other coordinates 0) and for $p \in P$, let f_{sp} denote the p coordinate of that restriction. If $f_s \neq 0$ for infinitely many s, an $x_s \in K$ can be chosen so that for at least one $p \in P$, $f_{sp}(x_s) = 1$. If $\bar{x}_s \in K^S$ denotes the element with x_s in coordinate s and 0 in all others the net of finite sums of $\bar{x}_s$ converges in the product topology to the element $x = (x_s)$. Thus the net of finite sums of $f(x_s)$ must also converge. But exactly as in convergence of ordinary infinite series this net can converge only if for every seminorm p and every $\epsilon > 0$, it is the case that $p(x_s) < \epsilon$ with at most finitely many exceptions. (The proof is the same, the finite partial sums cannot be a Cauchy net otherwise.)

Thus only finitely many f_s are non-zero. Now $f_s : K \longrightarrow P \cdot K$ is determined by $f_s(1)$ which involves finitely many $p \in P$. Thus in all only finitely many $f_{sp} \neq 0$. This shows that algebraically

$$(K^S, P \cdot K) \cong (S \times P) \cdot K .$$

As for the topology, we have seen already that every compact set in K^S is contained in a set of the form $\Pi \Delta(r_s)$ where r_s is a non-negative real number. A basic neighborhood of 0 in $P \cdot K$ is the set $\Gamma \Delta^\circ(t_p)$ where t_p is a positive real number or ∞. If $f : K^S \longrightarrow P \cdot K$ has components f_{sp} the $f_{sp} : K \longrightarrow K$ can be identified with elements of K. We have that

$$f(\Pi \Delta(r_s)) \subset \Gamma \Delta^\circ(t_p)$$

iff

$$(*) \qquad \sum_{s,p} f_{sp} \, r_s / t_p < 1 .$$

The reason is that $(\sum_s f_{sp} r_s \mid p \in P)$ must be able to be written as $(\lambda_p t_p)$ with $\lambda_p \in K$ and

$$\Sigma |\lambda_p| < 1 ,$$

which is easily seen to be equivalent to the condition $(*)$. Similarly the condition $(*)$ is equivalent to

$$(f_{sp}) \in \Gamma \Delta^\circ(t_p / r_s)$$

which is an open set in $(S\times P)\cdot K$. To go the other way, let $\Gamma\Delta^\circ(u_{sp})$ be an open set in $(S\times P)\cdot K$. That this is open in the function space topology is an easy consequence of the following.

(1.6) Lemma. Let f be a function from $\mathbb{N}\times\mathbb{N}$ - or a subset thereof - to the positive reals. Then there are functions $g : \mathbb{N}\longrightarrow \mathbb{R}$, $h : \mathbb{N}\longrightarrow \mathbb{R}$ such that

$$f(i,j) \geq g(i)h(j)$$

whenever (i,j) is in the domain of f .

Proof. By letting $f(i,j) = 1$ wherever f was hitherto undefined, we can suppose $f : \mathbb{N}\times\mathbb{N}\longrightarrow \mathbb{R}^+$. Let

$$g(n) = h(n) = \inf\{1, f(i,j) \mid i \leq n,\ j \leq n\} .$$

Then if, e.g., $i \leq j$, we have

$$1 \geq g(i)$$

$$f(i,j) \geq h(j)$$

$$f(i,j) \geq g(i)h(j)$$

Note: It is the failure of this lemma to hold for uncountable index sets that forces the restriction to finite and countable index sets.

(1.7) Now then, given an open set $\Gamma\Delta^\circ(u_{sp})$ in $(S\times P)\cdot K$, u_{sp} is a function on an index set $S\times P$ where each of S and P is either finite or countable. By the lemma we can find two sequences (r_s) and (t_p) such that

$$u_{sp} \geq r_s t_p .$$

This means that

$$\Gamma\Delta^\circ(u_{sp}) \supset \Gamma\Delta^\circ(r_s t_p)$$

and the latter is the set of doubly indexed sequences $f = (f_{sp})$ such that

$$f(\Pi\Delta(1/t_p)) \supset \Gamma\Delta^\circ(r_s)$$

and hence is a neighborhood of 0 in the function space.

(1.8) This decribes the bifunctor

$$\underline{C}^{op} \times \underline{D} \longrightarrow \underline{D}$$

which together with the duality between $\underline{C}$ and $\underline{D}$ describes a pre-$*$-autonomous situation.

That every object in $\underline{C}$ has a $\underline{D}$-representation is evident and the uniformity on the function space (C,D) is, by definition, a convergence uniformity.

(1.9) Proposition. The hypotheses and hence the conclusion of II.2.9 are satisfied (with $T = K$) .

Proof. The argument we used to show that any map $K^S \longrightarrow K$ is zero except for finitely many factors can be repeated to show that K is cosmall. Evidently any space in $\underline{A}$ is a topological vector space so that the injectivity of K follows from the Hahn-Banach theorem. The third hypothesis is evident.

(1.10) If $A \longrightarrow B$ is a proper closed embedding in $\underline{A}$ and $b \notin A$, $A+Kb/A \cong K$ which means that there is a non-zero map $A+Kb \longrightarrow K$ which is zero on A. The injectivity of K allows an extension of this to all of B. Thus $B^* \longrightarrow A^*$ is not injective.

It is evident that objects in $\underline{C}$ are complete. For the completeness of $\underline{D}$, see [Schaefer], II.6.2.

Thus the hypotheses required for our construction are satisfied.

The theory can also be worked out with $\underline{C} = \underline{D}$ = the category of finite sums (or products) of copies of K, equipped with the usual topology. The required details can be readily inferred from the above.

(1.11) Here is another interesting of a category of topological vector spaces. Again K stands for $\mathbb{R}$ or $\mathbb{C}$. A sequence (r_i) of elements of K is said to be *rapidly decreasing* if for all n, $\lim_{i\to\infty} i^n r_i = 0$. Since it is then also true that $\lim i^{n+2} r_i = 0$ is easily seen to imply that $\Sigma\, i^n |r_i| = 0$. A sequence (s_i) is said to be *slowly growing* if for some n, $|s_i| \le i^n$ for all sufficiently large i. Clearly for every rapidly decreasing $r = (r_i)$ and every slowly increasing $s = (s_i)$, the dot product $r\cdot s = \Sigma\, r_i s_i$ is absolutely convergent. We let D be the space of rapidly decreasing sequences, topologized by the seminorms $p_n(r) = \Sigma\, i^n |r_i|$ and C the space of slowly growing sequences topologized by uniform convergence on compact sets in D under the above pairing. Next I claim that C is the dual of D. First observe that the finite sequences are dense in D. For if p_n is the seminorm above and $r = (r_i) \epsilon D$, we know that $\lim i^{n+2} r_i = 0$ which means that for some m, $i > m$ $|i^{n+2} r_i| < 6\epsilon/\pi^2$ or that $|\sum_{m+1}^{\infty} i^n r_i| < \epsilon$. Thus r is approximated by the finite sequence which agrees with the first m terms of r and is 0 thereafter. Then a functional on D is determined by its value on the finite sequences which means it is determined by a sequence (s_i) by the formula $r_i \longmapsto \Sigma\, r_i s_i$. If, in fact, s_i is not slowly growing, we can choose a sequence $i(1) < i(2) < i(3) < \ldots$ such that $s_{i(j)} > i^j$. For having chosen $i(1),\ldots,i(j)$ we know that $s_i \le i^{j+1}$ is not satisfied for all sufficiently large i so there are arbitrarily large i for which it fails. Let $i(j+1)$ be the first one of these larger than $i(j)$. Then let $r_i = s_i^{-1}$ when $i = i(j)$ and 0 otherwise. Then $\Sigma\, r_i s_i$ diverges while $r_i < i^{-n-1}$ for all $i > i(n+1)$ implies $i^n r_i < i^{-1} \longrightarrow 0$ as $i \longrightarrow \infty$. This contradiction establishes that $(s_i) \epsilon C$.

(1.12) There is a certain class of spaces first isolated for study by Grothendieck called *nuclear spaces*. The usual definition is rather opaque but for our purpose the followed characterization of nuclear spaces conjectured by Grothendieck and established by T. and Y. Komura is more useful. Namely, a space is nuclear iff it is a subspace of a cartesian power of D. See [Pietsch] especially 11.1.1 for details. The topology on D is determined by a sequence of pseudo-norms and it is easily seen to be complete so that is a complete metric space, i.e. a Fréchet or F-space. Its dual C, with the bounded convergence topology, is a complete nuclear DF-space and both C and D are reflexive (in particular C has the topology of bounded convergence on D). See [Schaefer] II.7.1, corollary II.81; II.7.2., corollary 2; IV.5.6 and its first corollary; IV.6.1; IV.9.6.. Next I claim that $\mathrm{Hom}(C,D)$, equipped with the topology of uniform convergence on bounded sets, is isomorphic to D. In fact if $f : C \longrightarrow D$ is continuous, then it is continuous followed by each coordinate projection. So let f_j be the composite $C \longrightarrow D \longrightarrow \mathbb{R}$, the second map projection on the jth coordinate. Then f_j is a linear functional on C and hence represented by a sequence (r_{ij}) which for each j is rapidly decreasing. In order that for each $(s_i) \epsilon C$, the values $\Sigma s_i r_{ij} \epsilon D$ it is necessary and sufficient that the latter be a sequence rapidly decreasing in j. Although in principle this must be done for all (s_i) it is clearly sufficient to consider the test sequences (i^n). Thus (r_{ij}) represents a map iff for all n,m, we have $\lim i^n j^m r_{ij} = 0$. Clearly it is sufficient to restrict to $m = n$. To show the isomorphism with D we use the usual method of rearranging a double sequence into a sequence. So let $k(i,j)$ be the rearranging function. Since $k = k(i,j) \leq (i+j)^2$, $k^n |r_k| < (i+j)^{2n} |r_{ij}| \leq i^{2n} j^{2n} |r_{ij}| \longrightarrow 0$ as $k \longrightarrow \infty$. Conversely, $i < k(i,j)$ and $j < k(i,j)$ so $i^n j^n r_{ij} < k^{2n} r_k \longrightarrow 0$ as $i \rightarrow \infty$ or as $j \rightarrow \infty$. Thus the underlying vector space of $\mathrm{Hom}(C,D)$ is isomorphic to D. As for the topology, it follows from reflexivity that the polars of fundamental sequence of neighborhoods of 0 in D form a fundamental sequence of bounded sets in C. (See [Schaefer] Chapter IV, especially 5.2 in conjunction with the corollary of II.7.1.) What this means is the sets in C defined by $|s_i| < i^n$ form a fundamental system of bounded sets in C. An f represented by (r_{ij}) takes the set into the neighborhood $\{(t_j) \mid p_m(t_j) < \epsilon\}$ iff $\Sigma i^n j^m |r_{ij}| < \epsilon$. Thus the topology on $\mathrm{Hom}(C,D)$ is determined by these seminorms which are, as seen above, equivalent to the ones on D under the isomorphism.

(1.13) Let $\underline{C}$ be the full subcategory whose objects are finite direct sums of copies of K and C and $\underline{D}$ the full subcategory whose objects are finite direct sums (=products) of copies of K and D. We define the internal hom functor $(-,-): \underline{C}^{op} \times \underline{D} \longrightarrow \underline{D}$ so that $(K,K) = K$ and $(\Sigma_\psi C_\psi, \Pi_\omega D_\omega) = \Pi_{\psi,\omega}(C_\psi, D_\omega)$, the index sets being finite. The Hahn-Banach theorem implies that $\mathbb{R}$ is injective. It is also cosmall. In fact if $f : \Pi A_\psi \longrightarrow K$ is a functional, let $f_\psi = f|A_\psi$. Unless $f_\psi = 0$ there is an $a_\psi \epsilon A_\psi$ with $f(a_\psi) = 1$. If this happens infinitely often let $a = (a_\psi)$ with $a_\psi = 0$ whenever $f_\psi = 0$. The elements with a_ψ in finitely many coordinates and 0 in the remaining converge, in the product topology, to a. The value of f at such an element is the number of non-zero coordinates. Since this grows without bound, $f(a)$ cannot be defined. Thus $f_\psi = 0$ with only finitely many exceptions. If now $a = (a_\psi) \epsilon \Pi A_\psi$ has $f_\psi(a_\psi) = 0$ for all a, the same limit argument shows that $f(a) = 0$. Thus f factors through the projection on the finite product of those A_ψ for which $f_\psi \neq 0$. Hence the conditions of II.2.8 are satisfied. The condition (V) of III.4.2. is satisfied by exactly the same argument as in the preceding example (see 1.10). Hence all the hypotheses required for our main construction are satisfied. The category $\underline{A}$ consists, by the above mentioned result of T. and Y. Komura of exactly the nuclear spaces. As for ζ-complete spaces, we can examine one of its consequences fairly concretely here. The space C consists of slowly growing sequences. The subspace F of finite sequences is dense. In fact, if not there would be a non-zero linear functional on $C/c\ell(F)$ which means $\varphi \epsilon D$, $\varphi \neq 0$ but $\varphi | F = 0$. But φ is represented by a rapidly decreasing sequence and as soon as single coordinate is non-zero there is an $a \epsilon F$ with

$\varphi(a) \neq 0$. For $A \in \underline{A}$, a map $f : F \longrightarrow A$ is determined by a sequence $a_1, a_2, \ldots$ of elements of A . Here a_i is the image of the sequence with a 1 in the ith coordinate and 0 elsewhere. Continuity requires that for every seminorm p on A , the sequence $pa_1, pa_2, \ldots$ be extendable to all of C by the formula

$$pf(c) = pf((c_i)) = \Sigma\, p(c_i a_i) = \Sigma\, c_i pa_i$$

which means that (pa_i) must be a rapidly decreasing sequence of real numbers. Now say that a sequence (a_i) of elements of A is rapidly decreasing if the sequence (pa_i) is rapidly decreasing for all seminorms p . Then we require that whenever (a_i) is a rapidly decreasing sequence and (c_i) a slowly increasing sequence of real numbers, the sum $\Sigma\, c_i a_i$ converge. Since $(c_i a_i)$ is also rapidly decreasing, this can be replaced by the simpler hypothesis that every rapidly decreasing series converge. I must emphasize that this is only a consequence of ζ-completeness. It is tempting to conjecture that it is equivalent but there is no real evidence for that. If the conjecture were valid, it would follow that A is ζ-*-compact provided the sum of rapidly converging series of continuous functionals were always continuous.

2. Dualizing Modules.

(2.1) This example is in response to a question of Robert Raphael as to whether the theory of vector spaces over a discrete field had any natural generalization to modules over other commutative rings. If there is such a generalization, it seems likely that the dualizing object T will have to be injective and its dual - which is its endomorphism ring - must be the given ring. Technical considerations seem to require that T be finitely generated and a cogenerator. Accordingly let R be a commutative ring. We say that an R-module T is a *dualizing module* provided T is a finitely generated injective cogenerator and the canonical map $R \longrightarrow \mathrm{Hom}_R(T,T)$ is an isomorphism.

We leave till 2.10 the question of the existence of a dualizing module. If R is a field, it is clear that R itself is the unique dualizing module. The theory in that case reduces to that of vector spaces considered earlier.

(2.2) Now we let $\underline{V}$ be the category of all R-modules with its usual monoidal closed structure. The category Un $\underline{V}$ is simply that of topological R-modules. For $\underline{C}$ we take all modules of the form R^S with the product topology and for $\underline{D}$ all discrete modules of the form $S \cdot T$. Variations on the theory can be obtained by putting cardinality restrictions on S such as that it be finite or countable.

(2.3) If $D \in \underline{D}$, $D \cong S \cdot T$ and $\mathrm{Hom}(D,T) \cong \mathrm{Hom}(S \cdot T, T) \cong R^S$. We would like to define $D^* = R^S$ equipped with the product topology. But we must show that this is independent of the representation of D as a direct sum. Instead, define D^* to be Hom(D,T) equipped with the coarsest topology such that $D^* \longrightarrow R$ is continuous for all $T \longrightarrow D$. The coordinate injections $T \longrightarrow S \cdot T$ dualize to the projections $R^S \longrightarrow R$ so that this topology is at least as fine as the product topology. On the other hand, since T is finitely generated, any $T \longrightarrow S \cdot T$ factors through $F \cdot T$ for a finite subset $F \subset S$. When R^S and R^F have the product topology (the latter being discrete) the projection $R^S \longrightarrow R^F$ is continuous and hence so is $R^S \longrightarrow R^F \longrightarrow R$. Thus the topology on D^* is the product topology on R^S.

(2.4) If $C \cong R^S \in \underline{C}$, we see that any continuous $R^S \longrightarrow T$ must have an open submodule in its kernel. The product topology has a neighborhood base consisting of the kernels of maps $R^S \longrightarrow R^F$ where F is a finite subset of S. Thus $\text{Hom}(R^S,T)$ is the direct limit of $\text{Hom}(R^F,T) \cong \text{Hom}(F\cdot R,T) \cong \text{Hom}(R,T)^F \cong F\cdot\text{Hom}(R,T) \cong F\cdot T$ and the direct limit of the finite sums is just $S\cdot T$. Thus we define $C^* = S\cdot T$. This describes the duality between $\underline{C}$ and $\underline{D}$.

(2.5) Since T is a cogenerator, every $A\in\text{Un}\underline{V}$ has a $\underline{D}$-representation iff it is topologized by open submodules. Then $\underline{A}$ consists of these "linearly topologized" modules.

Proposition. T is injective with respect to the class of embeddings in $\underline{A}$.

Proof. Let $A \longrightarrow B$ be an embedding and $\varphi : A \longrightarrow T$ a map. Since T is discrete $A_0 = \ker\varphi$ is open in A, $A_0 = A\cap U$ where U is a neighborhood of 0 in B. Then $U \supset B_0$ where B_0 is an open submodule. I claim that $A_0 = A\cap(A_0+B_0)$. In fact if $a = a_0+b_0 \in A$ where $a_0\in A_0$, $b_0\in B_0$, then $b_0 = a-a_0 \in A$ while $b\in B_0 \subset U$ so $b\in A\cap U = A_0$. Thus $a = a_0+b_0 \in A_0$ as well. Thus we have

$$A/A_0 \longrightarrow B/(A_0+B_0)$$

is an injection and both modules are discrete. Since $A_0 = \ker\varphi$, φ induces $\varphi^\# : A/A_0 \longrightarrow T$. Then extension to $B/(A_0+B_0) \longrightarrow T$ now follows from the hypothesis that T is injective in $\underline{V}$.

(2.6) The hypotheses of II.2.9 are now satisfied. The cosmallness of T follows from the easily proved fact that a continuous map on a topological group is uniformly continuous together with I.2.12. The functor $\underline{C}^{op}\times\underline{D} \longrightarrow \underline{D}$ is very simply the hom-functor. For a continuous map $R^S \longrightarrow P\cdot T$ factors, by the previous remark, through R^F for some finite subset $F \subset S$. Since $R^F \cong F\cdot R$ is finitely generated a map to $P\cdot T$ factors through $G\cdot T \cong T^G$ for some finite $G \subset P$. Thus $\text{Hom}(R^S,P\cdot T) \cong \varinjlim \text{Hom}(F\cdot R,T^G) \cong \varinjlim \text{Hom}(R,T)^{F\times G} \cong \varinjlim (F\times G)\cdot T \cong (S\times P)\cdot T$ where the limits are taken over the finite subsets of S and P respectively. Thus we define (C,D) to be $\text{Hom}(C,D)$ with the usual structure of an R-module and the discrete topology.

(2.7) We can now verify the hypotheses of III.4.2. The first two are clear while the third is easy. In fact the required isomorphisms, such as

$$(R^S,(R^P,Q\cdot T)) \cong (R^Q,(R^P,S\cdot T))$$

are immediate when S,P and Q are finite and the general case follows by a limit argument as above. Since both sides are discrete, no topological question arises. Objects in $\underline{C}$ are powers of objects in $\underline{D}$ and hence have a $\underline{D}$-representation. The uniformity on (C,D) is a convergence uniformity, $\Phi(C)$ consisting of C alone (or of all subset of C).

Since the hypotheses of II.2.9 are satisfied, every object is prereflexive so III.4.2 (iv) is satisfied. The fifth hypothesis is an easy consequence of the fact that T is a cogenerator. In fact if $A \longrightarrow B$ is a proper embedding and $x\in B$, $x\notin A$, let B_0

be an open submodule of B such that $x+B_0$ does not meet A. But then $x \notin A+B_0$ which means the latter is a non-zero open submodule. Thus the non-zero discrete module $B/A+B_0$ admits a non-zero map to the cogenerator T. This is a map non-zero on B but 0 on A, as required.

That the objects in $\underline{C}$ and $\underline{D}$ are complete is evident and hence all the hypotheses are satisfied and the construction works.

(2.8) If R is noetherian, the full subcategory of finitely generated R-modules is a sub-$*$-autonomous category. In fact, T is finitely generated by hypothesis and if M and N are finitely generated, choose a surjection $F\cdot R \longrightarrow M$ with F finite. Then $\mathrm{Hom}(M,N)$ is a submodule of $\mathrm{Hom}(F\cdot R,N) \cong F\cdot N$. In that case it becomes natural to ask whether this is a *compact category* in the sense of Kelly, that is whether (M,N) is canonically isomorphic to $M^* \otimes N$. First off, by letting $M = N = R$ we see that it is necessary that $R \cong T$. Actually the question cannot be properly formulated without that hypothesis. Granting that, there are natural maps $M^* \otimes M \longrightarrow R$ and $R \longrightarrow (N,N)$ which compose to give $M^* \otimes M \longrightarrow (N,N)$. This transposes to $M^* \otimes N \longrightarrow (M,N)$ and what is really asked is whether this map is an isomorphism. It is easy to see that it is when R is a field and hence when R is semisimple (i.e. a finite product of fields). If R is not semisimple, there is a non-flat module (von Neumann regular self-injective rings are fields), and hence a finitely generated non-flat module (Tor commutes with $\varinjlim$), call it M. Then Steve Schamuel's elegant observation that $(M^*,-)$ is left exact while $M\otimes-$ is not settles the question. In fact if K is a field and $R = K[x]/(x^2)$, then it is easily seen that R is a dualizing module and that when K is made into an R-module via the obvious argumentation $R \longrightarrow K$, then the canonical map $K^* \otimes K \longrightarrow (K,K)$ is zero (although they are isomorphic).

(2.10) Now we turn to the question of the existence of a dualizing module. The only result I know is the following.

<u>Proposition</u>. Suppose K is a commutative ring with a dualizing module Q and R is a finitely generated K-projective K-algebra. Then for any constantly rank 1 finitely generated R-projective R-module P, $T = \mathrm{Hom}_K(P,Q)$ is a dualizing module for R.

Before beginning the proof let me observe that this requires beginning with a ring that has a dualizing module. Of course K might be a field. The result of this proposition applied in that case is that any finite dimensional commutative K-algebra has a dualizing module and may well have more than one. It can be shown that $\mathrm{Hom}_K(R,K) \cong R$ iff R is self injective.

Proof of the proposition. The proof that the T so defined is injective is standard. For if $M \longrightarrow N$ is an injection of R-modules, we have

$$\begin{array}{ccc} \mathrm{Hom}_R(N,\mathrm{Hom}_K(P,Q)) & \longrightarrow & \mathrm{Hom}_R(M,\mathrm{Hom}_K(P,Q)) \\ \downarrow \cong & & \downarrow \cong \\ \mathrm{Hom}_K(P\otimes_R N,Q) & \longrightarrow & \mathrm{Hom}_K(P\otimes_R M,Q) \end{array}$$

and with P R-projective $P\otimes_R M \longrightarrow P\otimes_R N$ is still an injection and since Q is K-injective it follows that the map is a surjection. Similarly, from $M \neq 0$, $\mathrm{Hom}_R(M,\mathrm{Hom}_K(R,Q)) \cong \mathrm{Hom}_K(M,Q)$ and the fact that Q is a cogenerator in K-modules, it follows that $\mathrm{Hom}_R(M,T) \neq 0$ so that T cogenerates R-modules. Since P is finitely generated as an R-module, there is a surjection $F\cdot R \longrightarrow P$ and since P is projective this map has a left inverse. Similarly R is a retract of $G\cdot K$ for some finite set G and thus as a K-module, P is a retract of $(F\times G)\cdot K$ whence $T = \mathrm{Hom}_K(P,Q)$ is a retract of $Q^{F\times G}$. Since Q is finitely generated as a K-module so is T. *A fortiori*, it is finitely generated as an R-module. The functor $M \longmapsto \mathrm{Hom}_K(\mathrm{Hom}_K(M,Q),Q))$ is finitely additive. The natural map

$$M \longrightarrow \mathrm{Hom}_K(\mathrm{Hom}_K(M,Q),Q)$$

is an isomorphism when $M = K$, by hypothesis, hence is when M is finite free and as well when it is finitely generated and projective. This is, in particular, true of P, considered as a K-module. Now

$$\begin{aligned}\mathrm{Hom}_R(T,T) &= \mathrm{Hom}_R(\mathrm{Hom}_K(P,Q),\mathrm{Hom}_K(P,Q))\\ &\cong \mathrm{Hom}_K(P\otimes_R\mathrm{Hom}_K(P,Q),Q)\\ &\cong \mathrm{Hom}_R(P,\mathrm{Hom}_K(\mathrm{Hom}_K(P,Q),Q))\\ &\cong \mathrm{Hom}_R(P,P)\quad .\end{aligned}$$

But P is locally isomorphic to R and hence the natural map $R \longrightarrow \mathrm{Hom}_R(P,D)$ is everywhere locally an isomorphism and hence is an isomorphism (its kernel and cokernel are everywhere locally zero).

3. Banach Spaces.

(3.1) Let $\underline{V}$ be the category of banach spaces and continuous linear maps of norm ≤ 1. It is known that this is a semivariety. The underlying functor assigns to each spare V its unit ball uV. The left adjoint assigns to each set S the banach space $\ell^1(S)$ of all functions $a : S \longrightarrow K$ (where K is the real or complex field) such that $\sum_{s\in S} |a_s|$ converges, with norm defined by $\|a\| = \Sigma|a_s|$. An algebra for the theory determined by this adjoint pair is a set closed under operations $(x_i) \longmapsto \Sigma\, \lambda_i x_i$ where (λ_i) is any finite or countable sequence of scalars for which $\Sigma\,|\lambda_i| \leq 1$. The operations must satisfy equations which look like double summation identities and are fairly obvious. Examples of algebras for this theory which are not (unit balls of) banach spaces are the *open* interval $(-1,1)$ as well as the quotient space $[-1,1]/(-1,1)$. This latter space is more accurately the coequalizer of the inclusion map and the zero map. It has three elements 1,-1 and the third which might be denoted 0, and represents the whole interior of the ball. Any operation not required by an identity to take the value 1 or -1, takes the value 0.

(3.2) Proposition. Let V be a banach space and $B \subset uV$ be topologically closed and invariant under the above theory. Then there is banach space W such that $B \cong W$.

Proof. Since uV is closed in V, so is B. Since V is complete, so is B. Now let W be the linear subspace of V generated by B. Since B is invariant under the operations it is convex and circled and hence determines a norm p on W by the formula

$$p(w) = \inf \{\lambda \mid w \in \lambda B\} .$$

If $w \in B$, $p(w) \le 1$, while if $p(w) \le 1$, $w \in (1+\epsilon)B$ for all $\epsilon > 0$ or $(1-\epsilon)w \in B$ for all $\epsilon > 0$. Since B is closed this implies $w \in B$. Thus B is closed in this norm. Now W is complete iff every set λB is since every Cauchy sequence is eventually in a λB. This is closed in V since B is. The uniformity induced by p on λB is finer than that induced by V. The uniform covers of λB are by translates of ϵB and these are closed in V since B is. Thus the completeness of λB in the p uniformity follows, by I.2.5, from its completeness in V.

(3.3) From this we see that the hypothesis of I.3.10 is satisfied. It is not clear what the uniform objects are. Here is an example of a preuniform object which is not uniform. First let I be the closed interval $[-1,1]$ with its usual uniformity. Let J be the same interval with the uniformity in which 1 and -1 are isolated while $(-1,1)$ has the uniformity inherited from the closed interval. This is a preuniform object because the value 1 (resp. -1) is the value of an operation iff all the genuine variables (i.e. all variables on which the operation genuinely varies) are 1 (resp. -1). A map $I \longrightarrow J$ preserves the algebra structure iff it is multiplication by a scalar of absolute value ≤ 1. Of these, only those of absolute value < 1 are uniform. Thus $\mathrm{Hom}(I,J)$ is $(-1,1)$ which does not belong to V.

(3.4) Fortunately, it is not necessary to describe $\mathrm{Un}\,\underline{V}$. It is sufficient to describe the full subcategories $\underline{C}$ and $\underline{D}$ and let $\underline{A}$ consist of all objects with a $\underline{D}$-representation. There are two natural choices for $\underline{C}$ and $\underline{D}$. The first is to take $\underline{C}$ and $\underline{D}$ to be finite dimensional banach spaces. Let A be a finite dimensional banach space with norm p and $a_1,\ldots,a_n$ be a linear basis. Let $\| \ \|$ denote the euclidean norm determined by $a_1,\ldots,a_n$, i.e.

$$\|\lambda_1 a_1+\ldots+\lambda_n a_n\| = (a_1^2+\ldots+a_n^2)^{\frac{1}{2}} .$$

Now let $\mu = \max(p(a_1),\ldots,p(a_n))$. First observe that from the ordinary inner product in R^n, we have

$$|(|\lambda_1|,\ldots,|\lambda_n|)\cdot(|1,\ldots,1)| \le \|\,|\lambda_1|,\ldots,|\lambda_n|)\|\ \|(1,\ldots,1)\|$$

$$|\lambda_1|+\ldots+|\lambda_n| \le \sqrt{n}\ (|\lambda_1|^2+\ldots+|\lambda_n|^2)^{\frac{1}{2}} .$$

Then

$$\begin{aligned} |p(\lambda_1 a_1+\ldots+\lambda_n a_n)| &\le |\lambda_1|p(a_1)+\ldots+|\lambda_n|p(a_n) \\ &\le \mu(|\lambda_1|+\ldots+|\lambda_n|) \\ &\le \mu\sqrt{n}(|\lambda_1|^2+\ldots+|\lambda_n|^2)^{\frac{1}{2}} \\ &= \mu\sqrt{n}\ \|\lambda_1 a_1+\ldots+\lambda_n a_n\| , \end{aligned}$$

so that p is bounded in the norm $\| \ \|$. Conversely, the $(n-1)$ sphere $\|a\| = 1$ is compact in a finite dimensional euclidean space and p never vanishes on it. Hence p takes on a minimum value there, say $1/M$, and it is clear that $\|a\| \le Mp(a)$ for all $a \in A$.

Thus each of the norms defines the same topology (or uniformity) on A .

(3.5) From this it follows every linear functional $A \longrightarrow R$ is continuous (but not necessarily a *map* in the category) and we let A^* be the set of them. We norm A^* by

$$p^*(f) = \sup \{ |f(a)| \mid p(a) \leq 1 \} .$$

We identify A^{**} with a under the usual identification of a finite dimensional space and its second dual. Let p^{**} denote the norm on A induced by p^* .

(3.6) Proposition. $p^{**} = p$.

Proof. For any $a \in A$, $|f(a/p(a))| \leq p^*(f)$ so that $|f(a)| \leq p^*(f)p(a)$. If $p^*(f) \leq 1$, $|f(a)| \leq p(a)$ and hence

$$p^{**}(a) = \sup \{ |f(a)| \mid p^*(f) \leq 1 \} \leq p(a) .$$

To go the other way, let $a \neq 0$. We may, without loss of generality, suppose $p(a) = 1$. The subspace generated at a certainly admits a functional f with $f(a) = 1$. The Hahn-Banach theorem ([Schaefer] , II.3.2.) guarantees that we can extend f to a functional defined on all of A such that $|f(b)| \leq p(b)$ for all $b \in A$. This implies that $p^*(f) \leq 1$. Then $p^{**}(a) \geq f(a) = 1 = p(a)$.

(3.7) This shows that $A \cong A^{**}$ and establishes the duality for finite dimensional spaces. An object of the category $\underline{A}$ is a set A which is the unit ball of a banach space and has the norm and the uniformity induced by a family of embeddings into unit balls of finite dimensional banach spaces. Note that the product uniformity does not coincide with the uniformity induced by the norm.

(3.8) Several observations may be made here. First off, a product of finite dimensional banach spaces, although not a banach space, is a topological vector space. In particular this means that the induced uniformity, restricted to the unit ball, has an extension to the whole space of such a nature that addition and scalar multiplication are uniformly continuous. This is assurdly not the case for the object J described in 3.3. There are two ways of extending this uniformity. The first is to use the uniformity induced by the product uniformity on the product of finite dimensional banach spaces. The second is to use the uniformity induced by the coarsest locally convex topology on the whole space such that every map to an arbitrary locally convex space which is continuous on the unit ball, is continuous. As long as this is done in the same way for every object in $\underline{A}$ it makes no difference which is chosen. The result is a category of "MT" (for mixed topology, see [Semadeni] ,[Wiweger]) spaces which is equivalent to the category $\underline{A}$. It is clear that the first approach is more in the spirit of our previous development. Thus we take $\underline{A}$ to be the category of spaces which are subspaces of products of discrete spaces with the norm and the topology induced.

It should be noted that if A and B have isomorphic (algebraically and topologically) unit balls they are isomorphic. For the topology on A is such that any map of A to a banach space, is continuous as soon as it is on the unit ball of A . Since B is embedded in a product of banach spaces, the same is true of B .

(3.9) As internal homfunctor $\underline{C}^{op} \times \underline{D} \longrightarrow \underline{D}$ we take (C,D) to be all linear maps between the spaces C and D . If p and q are the norm functions on C and D respectively, define the norm $(q/p)(f) = \sup\{qf(c) \mid p(c) \le 1\}$. This could also be described by

$$\sup\{qf(c/p(c)) \mid c \neq 0\}$$

except when $C = 0$. Using this it is easy to see that $\underline{C}$ and $\underline{D}$ form a pre $*$-autonomous situation. The only part of III.4.2 which must be demonstrated is that every object is prereflexive (the ζ-completeness here is a vacuous hypothesis). This becomes harder to verify as $\underline{C}$ and $\underline{D}$ grow larger for the dual has the same underlying $\underline{V}$ object but its uniformity becomes finer meaning it might admit more functionals. Accordingly we turn to the next example.

(3.10) We let $\underline{D}$ denote the category of banach spaces. The discussion of 3.8 goes through unchanged with finite dimensional banach spaces replaced by banach space. Thus we can consider $\underline{A}$ to be the category of those MT spaces whose topology and norm are determined by maps to banach spaces. Among them are the full subcategory $\underline{C}$ of these spaces which are locally convex and whose unit ball is compact (not, of course, in the norm but in that topology). Semademi shows ([Semademi]), that $\underline{C}^{op} \cong \underline{D}$. The duality may be described as follows. For $D \in \underline{D}$, D^* is the set of linear functionals on D topologized by pointwise (simple, weak) convergence and with the norm given by the same formula as in (3.5). The fact that the unit ball of D^* is compact is standard. Here is a proof. It is topologized as a subspace of K^D . If uD is the unit ball of D, a map in the unit ball of D^* takes uD to $I = uK$ and so that unit ball is topologized as a subspace of I^{uD} which is compact. It is in fact a closed subspace for any map $uD \longrightarrow I$ which preserves the *finitary* operations essentially convex linear combinations automatically induces a map $D \longrightarrow \mathbb{R}$. The preservation of a finitary operation involves only a finite number of coordinates and the set of maps preserving a given operation is easily seen to be closed. On the other hand, if $C \in \underline{C}$, let C^* be the set of continuous linear maps $C \longrightarrow K$. Each such map is bounded above on the compact unit ball of C and the least upper bound is the norm of the map. No additional topology is required on C^* . Before continuing, we require the following.

(3.11) Proposition. Let $C \in \underline{C}$ and $D \in \underline{D}$. Then the topology on the unit balls uC^* and uD^* is that of uniform convergence on compact sets bounded in norm.

Proof. For C that is clear since it is the topology of uniform convergence on the unit ball and its scalar multiples. So let $D \in \underline{D}$ and X be a compact subset. A zero neighborhood in the compact convergence topology is

$$\{\varphi \in D^* \mid \|\varphi\| \le 1 \text{ and } |\varphi(x)| < \epsilon \text{ for all } x \in X\}$$

where $\epsilon > 0$. Now there is a finite subset $x_1,\ldots,x_n$ such that $X \subset \bigcup (x_i + (\epsilon/2)uD)$ where uD is the unit ball of D . The set

$$\{\varphi \in D^* \mid \|\varphi\| \le 1 \text{ and } |\varphi(x_i)| < \epsilon/2,\ i=1,\ldots,n\}$$

is a neighborhood of 0 in the pointwise convergence topology. But if φ belongs to the latter set and $x \in D$, $x = x_i + (\epsilon/2)y$ for some y with $\|y\| \le 1$. Then

$|\varphi(x)| \le |\varphi(x_i)| + \epsilon/2 \ | \ \varphi(y)| < \epsilon/2 + \epsilon/2 = \epsilon$ and thus φ belongs to the former.

(3.12) The result of this is that if $C \longrightarrow D$ is a map the induced map $D^* \longrightarrow C^*$ is continuous on uD^* and hence on D^*. The same is true of a continuous linear function. For the topology on C is coarser than that of the norm, a fact which follows immediately from the representation in $\underline{D}$. Thus a continuous linear function $C \longrightarrow D$ is also continuous in norm and hence bounded. It follows that some scalar multiple of it is a map $C \longrightarrow D$, induces a map $D^* \rightarrow C^*$ and dividing by the scalar gives us again a continuous linear $D^* \longrightarrow C^*$.

(3.13) We now define (C,D) to be the space of continuous linear functions $C \longrightarrow D$ equipped with the usual sup-on-the-unit-ball norm. The previous paragraph implies that there is a map $(C,D) \longrightarrow (D^*, C^*)$. It is routine to see that that map preserves norm. Both $(C',(C,D))$ and $(C,(C',D))$ can be identified as the set of all bilinear maps $C \times C' \longrightarrow D$ which are continuous on the product of the unit balls. The fact that the usual hom/cartesian product adjointness works well when the domain spaces are compact ([Kelley], Chapter 7, theorem 5, p.223) implies the same here. The only thing left is to show that objects in $\underline{C}$ and $\underline{D}$ are reflexive.

(3.14) Proposition. Let $A \in \underline{A}$ and $a \in A$ be an element of norm > 1. Then there is a functional φ on A such that $\varphi(a) > 1$ and $\|\varphi\| \le 1$.

Proof. Since the unit ball of a banach space is closed so is the unit ball in a product of banach spaces. This property holds as well for a subspace of such a product. Thus the unit ball uA is closed in A and $a \notin uA$. Thus there is a convex circled neighborhood M of 0 such that $(a+M) \cap uA = \emptyset$. It follows that $(a+\frac{1}{2}M) \cap (uA+\frac{1}{2}M) = \emptyset$ so that $N = uA+\frac{1}{2}M$ is a neighborhood of 0 whose closure does not contain a. The gauge of N, defined by

$$p(b) = \inf \{\lambda | b \in \lambda N\} ,$$

is a seminorm on A with $p(b) \le \|b\|$ for all b and $p(a) > 1$. The functional φ defined on the one dimensional subspace generated by a $\varphi(a) = p(a)$ has, by the Hahn-Banach theorem, an extension to all of A for which $\varphi(b) \le p(b) \le \|b\|$.

(3.15) This means that no matter how A^* is topologized, as long as it bears the sup norm, the natural map $A \longrightarrow A^{**}$ is norm preserving. For given any $\epsilon > 0$, there is a $\varphi \in A^*$ of norm ≤ 1 such that $|\varphi(a/(1-\epsilon)\|a\|)| > 1$ which means that $|\varphi(a)| > (1-\epsilon)\|a\|$. Thus the

$$\sup \{ |\varphi(a)| \ | \ \|\varphi\| = 1\} \ge \|a\|$$

while the other direction is automatic. It follows that a, considered as a functional on A^*, has norm equal to $\|a\|$.

(3.16) This implies, in particular, that $A \longrightarrow A^{**}$ is an injection. Let $C \in \underline{C}$. The topology on C^{**} is that of simple convergence on C^*. Since each $\varphi \in C^*$ is continuous, it follows that $C \longrightarrow C^{**}$ is continuous in that topology. We will show that the image is dense. Suppose $f \in C^{**}$. A neighborhood of f is determined by a finite number $\varphi_1, \dots, \varphi_n$ of functionals on C and an $\epsilon > 0$ as

$$\{g \ || \ (g-f(\varphi_i)| < \epsilon\} .$$

Now $\ker \varphi_i$ is a subspace of codimension 1 so that $\ker \varphi_1 \cap \ldots \cap \ker \varphi_n$ has finite codimension. Let $A = C/\ker \varphi_1 \cap \ldots \cap \ker \varphi_n$. Then A^* is isomorphic to a finite dimensional subspace of C^*, (in fact, they are generated by $\varphi_1, \ldots, \varphi_n$). The restriction of f to A^* is, by (3.7), represented by an $a \in A$. If $c \in C$ is a preimage of a then the corresponding element of C^{**} lies in the neighborhood of f described above (in fact for any $\epsilon > 0$). Now suppose $\|f\| < 1$. Then $\|a\| = \|f|A^*\| \leq \|f\| < 1$ and so c may also be chosen to have norm < 1. Since uC is compact, its image is closed but every element of C^{**} of norm < 1 is in the closure of that image. Thus the image of uC is uC^{**} and that of C is C^{**}. Since uC is compact that map is homeomorphism on unit balls and hence $C \longrightarrow C^{**}$ is an isomorphism (see the discussion in (3.8)).

(3.17) Proposition. Let A be embedded in B. Then any functional $\varphi : A \longrightarrow K$ with $\|\varphi\| < 1$ has an extension to B of norm < 1.

Proof. Since K is complete and continuous maps are uniformly continuous, φ has an extension to the closure of A. Thus we may suppose A closed in which case so is $A_0 = \ker \varphi$. In that case A/A_0 is embedded in B/A_0 and A/A_0 is a space of dimension 1 (except in the trivial case that $\varphi = 0$) generated by an element a such that $\varphi(a) = 1$, whence $\|a\| > 1$. We know from 3.14 that there is a linear functional ψ on B/A_0 with $\psi(a) > 1$ and $\|\psi\| \leq 1$. The required functional is

$$b \longrightarrow \psi(b)/\psi(a) .$$

(3.18) Proposition. Every $A \in \underline{A}$ is quasireflexive.

Proof. We have shown that every $C \in \underline{C}$ is reflexive. Suppose $\{q_\omega : C_\omega \longrightarrow A\}$ is a family of maps $C_\omega \in \underline{C}$ such that the induced $A^* \longrightarrow \Pi C_\omega^*$ is an embedding. Suppose $f: A^* \longrightarrow \mathbb{R}$ is a functional with $\|f\| < 1$. Then we know that f has an extension $f^\#$ to a functional on ΠC_ω^* whose norm is still < 1. Let $f_\omega = f^\# | C_\omega^*$. If $\Sigma \|f_\omega\| > \|f^\#\|$ (in particular if more than countably many $f_\omega \neq 0$) we could find a finite set $\omega = 1, \ldots, n$ such $\|f_1\| + \ldots + \|f_n\| > \|f^\#\|$. Let $\Sigma = \|f_1\| + \ldots + \|f_n\| - \|f^\#\|$ and $\varphi_i \in C_i^*$ such that $\|\varphi_i\| = 1$ and $f_i(\varphi_i) > \|f_i\| - \epsilon/n$. Then let $(\varphi_\omega) \in \Pi C_\omega^*$ have φ_i in the ith coordinate and 0 elsewhere. We have

$$\begin{aligned} f(\varphi_\omega) &= f_1(\varphi_1) + \ldots + f_n(\varphi_n) \\ &> \|f_1\| + \|f_2\| + \ldots + \|f_n\| - \epsilon \\ &= \|f^\#\| \end{aligned}$$

while $\|(\varphi_\omega)\| = 1$, a contradiction. Now each $f_\omega \in C_\omega^*$ is repesented by an element $c_\omega \in C_\omega$ and of course $\Sigma \|c_\omega\|$ converges. Then with $a_\omega = g(c_\omega)$, $\Sigma \|a_\omega\| < 1$ so that Σa_ω converges (in norm, *a fortiori* in the topology) to an $a \in A$. If $\varphi \in A^*$, $f(\varphi) = f^\#((\varphi g_\omega))$ $= \Sigma \varphi g_\omega(c_\omega) = \varphi(\Sigma g_\omega c_\omega) = \varphi(a)$. The next to last equality follows from the fact that φ is additive and continuous.

(3.19) Now with $D \in \underline{D}$, both D and D^{**} are banach spaces, the map between them is bijective and preserves norm (see 3.15) and is thus an isomorphism. This finishes the proof that $\underline{C}$ and $\underline{D}$ constitute a pre-*-autonomous situation.

We should now verify III.4.2(v). Unfortunately it is not quite true in the form stated. The condition is satisfied for norm-preserving embeddings (use the same argument as in 1.10). The way out is to let $\tau(C_1,C_2)$ have the norm of $|C_1|\otimes|C_2|$ and the topology of $(C_1,C_2^*)^*$. Then the map constructed to the Corollary of III.27 will necessarily be continuous and can be proved *ad hoc* to be norm non-increasing. To see this, we begin with

(3.19) Proposition. For any $A,B\in\underline{A}$, the natural embedding

$$\underline{A}(A,B) \longrightarrow \underline{A}(B^*,A^*)$$

is norm preserving.

Proof. If $f : A \longrightarrow B$,

$$\|f\| = \sup\{\|f(a)\| \,|\, \|a\| = 1\} .$$

Suppose $\|f\| = 1$. Then for any $\epsilon > 0$, there is an $a\epsilon A$ such that $\|a\| = 1$ and $\|f(a)\| > 1-\epsilon/2$. There is a linear functional β on B such that $\|\beta\| = 1$ and $|\beta f(a)| > 1-\epsilon$. But then

$$\|a\| = 1, \; \|f^*\beta(a)\| > -\epsilon$$

which implies that

$$\|f^*\beta\| > 1-\epsilon .$$

Since this is true for all $\epsilon > 0$ and $\|\beta\| = 1$,

$$\|f^*\| \geq 1 .$$

Since $\|f^*\| \leq \|f\| = 1$, it follows that

$$\|f^*\| = 1 .$$

Corollary. The natural embedding

$$\underline{V}(A,B) \longrightarrow \underline{V}(B^*,A^*)$$

preserves norm.

(3.20) Now we suppose that

$$f : C_1 \longrightarrow \underline{A}(C_2,A)$$

has norm 1 and that $\tilde{f} : (C_1,C_2^*)^* \longrightarrow A$ is the map induced by the Corollary to III.2.7. We must show $\tilde{f}$ has norm ≤ 1 . The map

$$f^\# : C_1 \longrightarrow A(C_2,A) \longrightarrow \underline{A}(A^*,C_2^*)$$

still has norm 1 and $\tilde{f}$ has the same norm as

$$\tilde{f}^* : A^* \longrightarrow (C_1,C_2^*) .$$

Now if $c_1\epsilon C_1$, $\alpha\epsilon A^*$, $c_2\epsilon C_2$ are such that $\|c_1\| = \|\alpha\| = \|\gamma_2\| = 1$, then

$$\|f^\#(c_1)\| \leq 1$$

which implies that

$$\|f^\#(c_1)(\alpha)\| \leq 1$$

and hence that

$$|f^\#(c_1)(\alpha)(c_2)| \leq 1 .$$

But $f^{\#}(C_1)(\alpha)(C_2) = \tilde{f}(\alpha)(C_1)(C_2)$ and since α, C_1, C_2 are arbitrary elements of norm 1, the inequality

$$|\tilde{f}(\alpha)(C_1)(C_2)| \leq 1$$

successively implies

$$\|\tilde{f}(\alpha)(C_1)\| \leq 1$$

$$\|\tilde{f}(\alpha)\| \leq 1$$

$$\|\tilde{f}\| \leq 1$$

which establishes what we need. Since no other use was made of III.4.2(v), we may ignore that condition here.

The objects in $\underline{D}$ are banach spaces and hence are complete. As for $\underline{C}$, recall that the "underlying set" is the unit ball and for spaces in $\underline{C}$, this is compact. Thus the requisites for our construction are present.

4. Modules over a Hopf Algebra.

(4.1) Let H be a cocommutative hopf algebra over a field K. This means that H is a vector space over K, equipped with K-linear maps

$$\eta : K \longrightarrow H \; ; \; \epsilon : H \longrightarrow K$$

$$\mu : H \otimes H \longrightarrow H \; ; \; \delta : H \longrightarrow H \otimes H$$

$$\iota : H \longrightarrow H$$

such that ϵ and δ determine a cocommutative coalgebra and η, μ, ι a group object in the category of cocommutative coalgebras.

(4.2) The best known example is the group algebra $K[G]$ of a group. Here $\epsilon(g) = 1$, $\delta(g) = g \otimes g$ and $\iota(g) = g^{-1}$ for $g \in G$. Note that the algebra is commutative iff G is. The second best known example is the enveloping algebra L^e of a lie algebra L. The operations are the unique algebra homomorphisms for which $\epsilon(\ell) = 0$, $\delta(\ell) = 1 \otimes \ell + \ell \otimes 1$ and $\iota(\ell) = -\ell$ for $\ell \in L$. Most of the many features which are common to the theory of groups and of lie algebras belong in fact to the theory of hopf algebras.

(4.3) If H is a hopf algebra we understand by an H-module a module for the associative algebra whose multiplication is given by μ. A morphism $f : M \longrightarrow N$ of H-modules is simply a module homomorphism. We let $\underline{H}$ denote the category of H-modules.

(4.4) We adopt Sweedler's notation and write for $h \in H$,

$$\delta h = \sum_h h_{(1)} \otimes h_{(2)}$$

although in practice, the index on the summation is usually omitted.

If M is a module, let $\mu_M : H \otimes M \longrightarrow M$ denote the action of H on M.

(4.5) Let $\underline{H}$ denote the category of H-modules and morphisms described above. If $M, N \in \underline{H}$, let $\underline{H}(M,N)$ denote the abelian group $\mathrm{Hom}(M,N)$ equipped with the H-action defined by

$$(hf)(m) = \sum h_{(1)} f(\iota h_{(2)} m) \; .$$

In the case of a group G, this formula amounts to

$$(xf)(m) = xf(x^{-1}m), \; x\epsilon G$$

which is more-or-less standard while for a lie algebra L, it becomes the "bracket"

$$(xf)(m) = xf(m) - f(xm), \; x\epsilon L .$$

(4.6) Proposition. Let M,N in $\underline{H}$. A K-linear map $f : M \longrightarrow N$ is a homomorphism of H-modules iff for all $h\epsilon H$, $m\epsilon M$,

$$(\epsilon h)f(m) = \Sigma\, h_{(1)}f(\iota h_{(2)}m) .$$

Proof. To motivate the proof we consider that case $H = K[G]$ where G is a group. In that case, the formula above becomes, for all $x\epsilon G$,

$$f(m) = xf(x^{-1}m)$$

which is evidently equivalent to

$$x^{-1}f(m) = f(x^{-1}m)$$

and since x^{-1} is arbitrary, this means

$$xf(m) = f(xm)$$

for all $x\epsilon G$. The proof below is a translation of this into diagrammatic language. It is worth mentioning that a purely combinatorial argument would seem nearly impossible. The proof is gotten by juxtaposing commutative diagrams. Since ι is an involution of the coalgebra structure $\epsilon\iota = \epsilon$, $\delta\iota = \delta$. If we replace h by ιh the formula above becomes

$$(\epsilon h)f(m) = (\Sigma\iota h)f(m) = \Sigma\iota h_{(1)}f(h_{(2)}m) .$$

This can be written

(*) $$\mu_N . H\otimes f . H\otimes\mu_M . \iota\otimes H\otimes M . \delta\otimes M = f . \epsilon\otimes M$$

where we use the canonical isomorphism to identify $K\otimes M$ with M. Then we have a commutative diagram

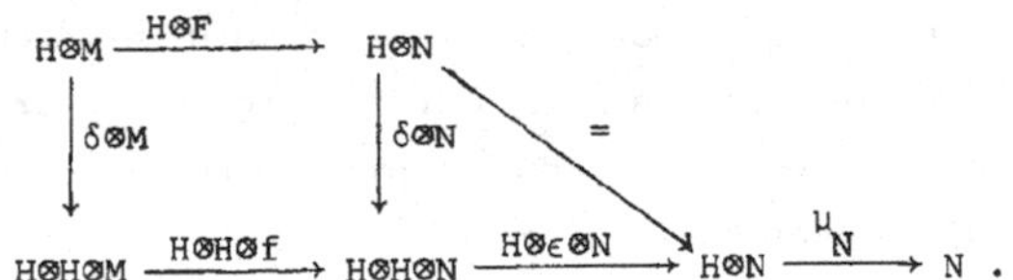

The left-lower sequence of this diagram is the upper-right of next. The polygon marked $H\otimes(*)$ is exactly that, (*) being the formula labelled above.

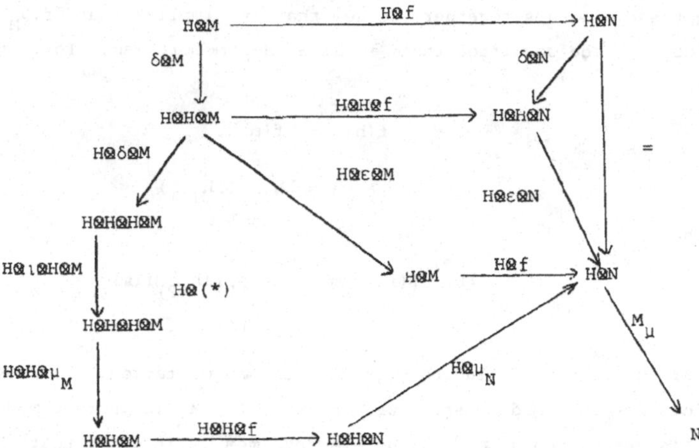

Then we have a commutative diagram

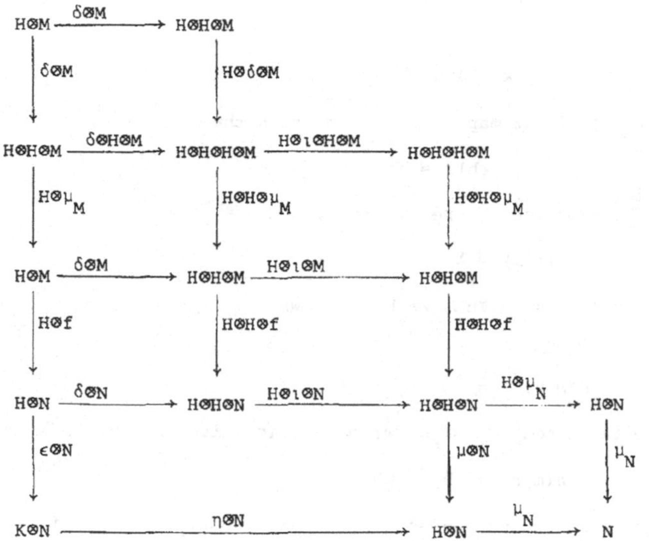

Finally we have a commutative diagram

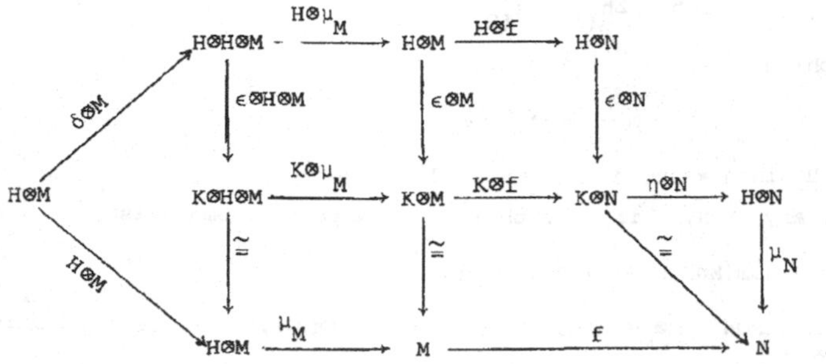

Putting these diagrams together, we see that (*) implies that $f.\mu_M = \mu_N.H\otimes f$. This is, of course, the condition that f be an H-homomorphism. To go the other way the equations

$$f(hm) = hf(m) ,$$

$$\epsilon h = \Sigma h_{(1)}(ih_{(2)})$$

imply that

$$\Sigma h_{(1)}f(ih_{(2)}m) = \Sigma h_{(1)}(ih_{(2)})f(m)$$
$$= (\epsilon h)f(m) .$$

(4.7) If H is a hopf algebra over K , the augmentation $\epsilon : H \longrightarrow K$ determines in an obvious way an H-module structure on K . If M is another module, a K-linear map $K \longrightarrow M$ is determined uniquely by an element $m \in M$. In order that the map be a morphism in the category the element m must satisfy, for all $h \in H$,

$$(\epsilon h)m = hm .$$

In particular, a map

$$K \longrightarrow \underline{H}(M,N)$$

is determined uniquely by a K-linear map $f : M \longrightarrow N$ such that

$$(\epsilon h)f = hf$$

which means, in view of the H-action defined in (4.4),

$$(\epsilon h)(fm) = \Sigma h_{(1)}f(ih_{(2)}m) ,$$

i.e. that f is an H-homomorphism. Thus we have shown,

<u>Proposition</u>. For any M,N in $\underline{H}$,

$$\mathrm{Hom}(M,N) \cong \mathrm{Hom}(K,\underline{H}(M,N)) .$$

(4.8) Define $M\otimes N$ to be the K-module $M\otimes_K N$ with H-action given by

$$h(m\otimes n) = \Sigma h_{(1)}m\otimes h_{(2)}n .$$

Since the algebra is cocommutative and coassociative, it is easy to see that this gives a symmetric, associative tensor. The counit identity

$$h = \Sigma h_{(1)}(\epsilon h_{(2)})$$

gives rise to an isomorphism

$$M \xrightarrow{\cong} M\otimes K$$

so that this $\otimes$ makes $\underline{H}$ into a symmetric monoidal category.

(4.9) <u>Proposition</u>. For any M,N,P in $\underline{H}$, there is a natural isomorphism,

$$\mathrm{Hom}(M\otimes N,P) \longrightarrow \mathrm{Hom}(M,\underline{H}(N,P)) .$$

Proof. Since as a K-module, $M\otimes N = M\otimes_K N$ and $\underline{H}(M,N) = \mathrm{Hom}_K(M,N)$, there is a canonical isomorphism,

$$\mathrm{Hom}_K(M\otimes N,P) \longrightarrow \mathrm{Hom}_K(M,\underline{H}(N,P)) .$$

Thus the only thing to check is that a morphism on one side corresponds to a morphism on the other. If $f : M\otimes N \longrightarrow P$ is a K-linear map it is a morphism iff it satisfies, for all $h\in H$, $m\in M$, $n\in N$,

$$(\epsilon h)f(m\otimes n) = \Sigma\, h_{(1)}f(\iota h_{(2)}(m\otimes n))$$

(i) $$= \Sigma\, h_{(1)}f(\iota h_{(2)(1)}m\otimes\iota h_{(2)(2)}n) .$$

The corresponding map $\tilde{f} : M \longrightarrow \underline{H}(N,P)$, defined by

$$\tilde{f}(m)(n) = f(m\otimes n)$$

must satisfy for all $h\in H$, $m\in M$

$$(\epsilon h)(\tilde{f}m) = \Sigma\, h_{(1)}\tilde{f}(\iota h_{(2)}m)$$

which means, for all $n\in N$,

$$(\epsilon h)(\tilde{f}m)(n) = \Sigma\, h_{(1)}\tilde{f}(\iota h_{(2)}m)(n)$$

$$(\epsilon h)f(m\otimes n) = \Sigma\, h_{(1)(1)}\tilde{f}(\iota h_{(2)}m)(\iota h_{(1)(2)}n)$$

(ii) $$= \Sigma\, h_{(1)(1)}f(\iota h_{(2)}m\otimes\iota h_{(1)(2)}n) .$$

To see that these are equal, let μ_M, μ_N, μ_P denote the maps describing the action of H on M,N,P, respectively (e.g. $\mu_M : H\otimes M \longrightarrow M$). Also let $\sigma : H\otimes H \longrightarrow H\otimes H$ denote the interchange isomorphism. From the cocommutativity and coassociativity, it follows that the diagram

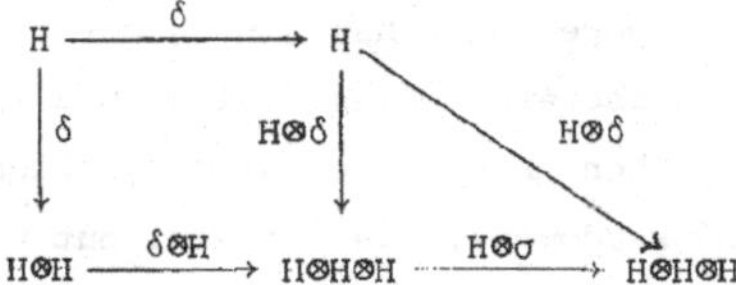

commutes. This implies that for $h\in H$

$$\Sigma\, h_{(1)}\otimes h_{(2)(1)}\otimes h_{(2)(2)} = \Sigma\, h_{(1)(1)}\otimes h_{(2)}\otimes h_{(1)(2)} .$$

By applying $H\otimes\iota\otimes\iota$, tensoring with $m\otimes n$ and applying an interchange, we conclude that

$$\Sigma h_{(1)}\otimes\iota h_{(2)(1)}\otimes m\otimes\iota h_{(2)(2)}\otimes n = \Sigma h_{(1)(1)}\otimes\iota h_{(2)}\otimes m\otimes\iota h_{(1)(2)}\otimes n .$$

Now apply $\mu_P.H\otimes f.H\otimes\mu_M\otimes\mu_N$ and the required identity of (i) and (ii) results.

(4.10) Theorem. $\underline{H}$ is an autonomous category.

Note that the set underlying $\underline{H}(M,N)$ is $\mathrm{Hom}_K(M,N)$, not $\mathrm{Hom}(M,N)$. Thus we have an example which features pseudomaps distinct from maps. In the development in chapter III it may have appeared that the pseudomaps should have been taken as the primary category. Although that point of view would have simplified the language, it would have led in the current instance to considering the category whose objects are H-modules but whose maps are just K-linear.

(4.11) Until now we have not used the fact that K is a field. Thus the preceding results are valid for arbitrary hopf algebras. What follows will require that K be a field although it would likely be sufficient that K have a dualizing module.

(4.12) It is evident that $\underline{H}$ is a variety as well as closed monoidal category. It should be noted that the theory of H is not in general commutative (the theory of K[G]-modules in commutative iff G is) and even if it is, the closed structure is not the canonical one. This is reflected in the existence of pseudomaps distinct from maps. The pseudomaps $M \longrightarrow N$ - i.e. the elements of $\underline{H}(M,N)$ - are the K-linear maps while the maps are the H-homomorphisms.

Since the category $\underline{H}$ is abelian, the category Un $\underline{H}$ is equivalent to topological H-modules. We let $\underline{D}$ be the discrete H-modules and $\underline{C}$ be the category of topological H-modules which are, as K-vector spaces, linearly compact. The dualizing object is K equipped with the H structure induced by ϵ. As usual we let $\underline{A}$ denote the subcategory of objects which have a $\underline{D}$-representation. Evidently the vector space underlying any such object is a topological K-vector space with enough discrete representations. We also know that if $M \in \underline{A}$, the space underlying $\underline{M}(|M|,K)$ is the same as the dual space of the vector space underlying $\underline{M}$. The same is evidently true of $M^* = \underline{A}(M,A)$. Thus the linear transformation underlying the canonical map

$$|M| \longrightarrow \underline{H}(M^*,K)$$

is an isomorphism and hence this map is an isomorphism. This shows that every object is quasi-reflexive.

To show that objects in $\underline{C}$ have a $\underline{D}$-representation, it is sufficient to observe that every topological K-vector space which has a neighborhood basis at 0 of open subspaces, has a representation by discrete spaces. If such a space is the vector space underlying an object of Un $\underline{H}$, then that object already belongs to $\underline{A}$. For it is sufficient to find a family of pseudomaps, i.e. linear, but not necessarily H-linear, maps. For that, you can take the maps to discrete spaces which can be given H-module structures as direct sums of copies of K. In particular this holds for all $C \in \underline{C}$. The pre-$*$-autonomous structure is evident. The module of continuous maps of $C \longrightarrow D$ is topologized discretely and given the H-structure described in (4.5). The continuity required of the H action insures that hf is continuous when f is. The remaining parts of III.4.2 are immediate. Recall, in showing the fifth, that the vector spaces underlying C^* and A^* are the continuous K-linear functionals on C and A, respectively. Thus the general theory applies here.

5. Topological Abelian Groups.

(5.1) In this section we let $\underline{V}$ be the category of abelian groups. We can and will, as observed at the beginning of this chapter, identify Un $\underline{V}$ with Top $\underline{V}$, the category of topological abelian groups.

Let A be an abelian group. By a seminorm on A is meant a function

$$p : A \longrightarrow \mathbb{R}$$

Such that

i) $p(0) = 0$;

ii) $p(a-a') \leqslant p(a) + p(a')$, for all $a, a' \in A$.

Trivial consequences are

iii) $p(a) \leqslant 0$ (take a=a' in (ii)) ;

iv) $p(a) = p(-a)$ (take a'=0 in (ii)) ;

v) $p(a+a') \leqslant p(a) + p(a')$;

vi) $|p(a) - p(a')| \leqslant p(a-a')$.

(5.2) An invariant pseudometric on A is a pseudometric d such that

i) $d(a,a') = d(a+a'', a'+a'')$, for all $a,a',a'' \in A$.

Trivial consequences are

ii) $d(a,a') = d(-a,-a')$;

iii) $d(a-a') = d(a-a'', a'-a'')$.

(To prove (ii), $d(-a,-a') = d(0, a-a') = d(a-a',0) = d(a,a')$.)

(5.3) Let p be a seminorm on A . Define a function

$$\alpha(p) : A \times A \longrightarrow \mathbb{R}$$

by

$$(a,a') = p(a-a') .$$

Then for $a,a',a'' \in A$,

$$\begin{aligned} \alpha(p)(a,a') &= p(a-a') \\ &\leqslant p(a-a'') + p(a''-a') \\ &= \alpha\ (p)(a,a'') + \alpha(p)(a'',a') \end{aligned}$$

so that $\alpha(p)$ is a pseudometric. Also,

$$\begin{aligned} \alpha(p)(a,a') &= p(a-a') \\ &= p(a+a''-a''-a') \\ &= p[a+a''-(a''+a')] \\ &= \alpha(p)(a+a', a'+a'') \end{aligned}$$

and we see that $\alpha(p)$ is an invariant pseudometric.

Let d be an invariant seminorm on A and define a function

$$\beta(d) : A \longrightarrow \mathbb{R}$$

by

$$\beta d(a) = d(a,0) .$$

Then

$$\beta d(0) = d(0,0) = 0$$

$$\begin{aligned}\beta d(a-a') &= d(a-a',\ 0)\\ &= d(a-a'+a',\ 0+a')\\ &= d(a,a')\\ &\le d(a,0) + d(0,a')\\ &= d(a,0) + d(a',0)\\ &= \beta d(a) + \beta d(a')\end{aligned}$$

so that βd is a seminorm.

(5.4) <u>Proposition</u>. The correspondences

$$p \longrightarrow \alpha(p)$$

$$d \longrightarrow \beta(d)$$

determine a 1-1 correspondence between seminorms on A and invariant pseudometrics on A.

Proof. We have

$$\begin{aligned}\alpha\beta d(a,a') &= \beta d(a-a')\\ &= d(a-a',\ 0)\\ &= d(a-a'+a',\ a')\\ &= d(a,a') \quad .\end{aligned}$$

Also,

$$\begin{aligned}\beta\alpha p(a) &= \alpha p(a,0)\\ &= p(a-0)\\ &= p(a) \quad .\end{aligned}$$

Thus α and β are inverse isomorphisms.

(5.5) <u>Proposition</u>. Let A be a topological group and p a seminorm. Then the following are equivalent:

i) p is continuous ;

ii) p is uniformly continuous ;

iii) $\alpha(p)$ is continuous ;

iv) $\alpha(p)$ is uniformly continuous .

Proof. If p is continuous (resp. uniformly continuous) $\alpha(p)$ is the composite

$$A \times A \xrightarrow{-} A \xrightarrow{p} \mathbb{R}$$

in which the first map, subtraction, is uniformly continuous, so the composite is. If $\alpha(p)$ is continuous (resp. uniformly continuous), $p = \beta\alpha(p)$ is the composite

$$A \longrightarrow A \times A \xrightarrow{\alpha(p)} \mathbb{R} \ .$$

The components of the first map are the identity and 0 and it is uniformly continuous. Finally, suppose that p is continuous. For $\varepsilon > 0$, choose a neighborhood M of 0 such that

$$a \in M \text{ implies } p(a) < \varepsilon \ .$$

Then for $a \in A$, $a' \in a+M$ implies

$$|p(a) - p(a')| \le p(a-a') < \varepsilon \ .$$

(5.6) Let A be a topological abelian group and M be a neighborhood of M. Then so is $-M$ and $M_0 = M \cap (-M)$. The set M_0 is a <u>symmetric</u> (i.e. $M_0 = -M_0$) neighborhood of 0 contained in M . We may inductively choose a sequence

$$M_1, M_2, M_3, \ldots .$$

of symmetric neighborhoods of 0 such that

$$M_n + M_n \subset M_{n-1}$$

for all $n > 0$.

If α is a finite set of strictly positive integers, let

$$\lambda(\alpha) = \Sigma\{2^{-i} | i \in \alpha\}$$

be the corresponding dyadic rational number.
If λ is a positive dyadic rational, define a subset $M(\lambda)$ of A by

$$M(\lambda) = A , \quad \text{if} \quad \lambda > 1$$

$$M(1) = M_0$$

$$M(\lambda(\alpha)) = \Sigma\{M_i | i \in \alpha\} \quad .$$

(5.5) <u>Proposition</u>. For any dyadic rationals μ and ν ,

$$M(\mu) + M(\nu) \subset M(\mu+\nu) \quad .$$

Proof. If $\mu+\nu > 1$, the conclusion is evident. If $\mu=1$ and $\nu=0$ or vice versa, the conclusion is also evident. Thus we may suppose

$$\mu > 1 \quad ; \quad \nu > 1 \quad .$$

This means there are finite sets of integers α and β such that

$$\lambda(\alpha) = \mu \quad ; \quad \lambda(\beta) = \nu \quad .$$

The proof is by a double induction; first on the cardinality of $\alpha \cap \beta$ and second on the smallest integer in $\alpha \cap \beta$. Thus we begin by supposing that

$$\alpha \cap \beta = \emptyset \ .$$

In that case, we have

$$\begin{aligned} \lambda(\alpha \cup \beta) &= \Sigma\{2^{-i} | i \in \alpha \cup \beta\} \\ &= \Sigma\{2^{-i} | i \in \alpha\} + \Sigma\{2^{-i} | i \in \beta\} \end{aligned}$$

since the union is disjoint. Thus

$$\lambda(\alpha \cup \beta) = \lambda(\alpha) + (\beta) \quad .$$

Similarly

$$\begin{aligned} M(\mu+\nu) &= M(\lambda(\alpha) + \lambda(\beta)) \\ &= M(\lambda(\alpha \cup \beta)) \\ &= \Sigma\{M_i | i \in \alpha \cup \beta\} \\ &= \Sigma\{M_i | i \in \alpha\} + \Sigma\{M_i | i \in \beta\} \\ &= M(\lambda(\alpha)) + M(\lambda(\beta)) \\ &= M(\mu) + M(\nu) \quad . \end{aligned}$$

Next we suppose that $\alpha \cap \beta \neq 0$ and that the conclusion is valid whenever

$$\mu = \lambda(\alpha') \quad ; \quad \nu = \lambda(\beta')$$

and $\alpha' \cap \beta'$ has fewer elements than $\alpha \cap \beta$ and that it is also valid when $\alpha' \cap \beta'$ has the same number of elements but the least integer of $\alpha' \cap \beta'$ is smaller than that of $\alpha \cap \beta$.

Now let i be the smallest element of $\alpha \cap \beta$. Then if $i = 1$,

$$\mu \geq \frac{1}{2} \ , \qquad \nu \geq \frac{1}{2}$$

and the only possibility consistent with

$$\mu + \nu \leq 1$$

is

$$\mu = \nu = \frac{1}{2}$$

in which case the result follows from

$$M_1 + M_1 \subset M_0 \ .$$

Now suppose $i > 1$. Since

$$i - 1 \not\subset \alpha \cap \beta$$

it is not in at least one of them. Suppose, say, that $i\text{-}1 \notin \alpha$. In that case, let

$$\alpha' \ = \ \alpha - \{i\} \cup \{i - 1\}$$

$$\beta' \ = \ \beta - \{i\}.$$

Then if $i\text{-}1 \notin \beta$, $\alpha' \cap \beta'$ has fewer elements than $\alpha \cap \beta$ while if $i\text{-}1 \in \beta$, they have the same number of elements but the least element of $\alpha' \cap \beta'$ is $i\text{-}1$ which is smaller than that of $\alpha \cap \beta$. In either case, our inductive hypothesis implies

$$M(\lambda(\alpha')) + M(\lambda(\beta')) \subset M(\lambda(\alpha') + \lambda(\beta')) \ .$$

Evidently

$$\lambda(\alpha') \ = \ \lambda(\alpha) - 2^{-i} + 2^{-i+1}$$

$$\lambda(\beta') \ = \ \lambda(\beta) - 2^{-i}$$

so that

$$\lambda(\alpha') + \lambda(\beta') \ = \ \lambda(\alpha) + \lambda(\beta) \ .$$

Finally,

$$\begin{aligned} M(\lambda(\alpha)) + M(\lambda(\beta)) &= \Sigma\{M_j | j \in \alpha\} + \Sigma\{M_j | j \in \beta\} \\ &= \Sigma\{M_j | j \in \alpha-\{i\}\} + M_i + \Sigma\{M_j | j \in \beta-\{i\}\} + M_i \\ &\subset \Sigma\{M_j | j \in \alpha-\{i\}\} + M_{i-1} + \Sigma\{M_j | j \in \beta'\} \\ &= \Sigma\{M_j | j \in \alpha'\} + \Sigma\{M_j | j \in \beta'\} \\ &= M(\lambda(\alpha')) + M(\lambda(\beta')) \\ &\subset M(\lambda(\alpha') + \lambda(\beta')) = M(\lambda(\alpha) + \lambda(\beta)) \ . \end{aligned}$$

Corollary. If $\mu \leq \nu$ are dyadic rationals, then

$$M(\mu) \subset M(\nu) \ .$$

Proof. For the difference of two dyadic rationals is one so that

$$M(\mu) \subset M(\mu) + M(\nu-\mu) \subset M(\nu) \ .$$

(5.6) Proposition. The topology (resp. uniformity) of any topological abelian group is given by a family of continuous seminorms (resp. uniform invariant pseudometrics).

Proof. Let A be a topological group and M a neighborhood. There is then a family

$$\{M(\lambda) \mid \lambda \text{ a dyadic rational }\}$$

of symmetric neighborhoods of 0 such that

$$M(1) \subset M$$

$$M(\lambda) + M(\mu) \subset M(\lambda+\mu)$$

and whenever $\lambda \leq \mu$,

$$M(\lambda) \subset M(\mu) \ .$$

Define a function

$$p : A \longrightarrow \mathbb{R}$$

by

$$p(a) = \inf\{\lambda | a \in M(\lambda)\}$$

as λ runs over all the positive dyadic rationals.

(5.7) Proposition. The function p has the following properties.

(i) $p(0) = 0$;

(ii) $p(a) = p(-a)$;

(iii) $p(a-a') = p(a) + p(a')$;

(iv) p is continuous ;

(v) $p(a) \leq 1$ implies $a \in M$.

Proof. (i) is evident from the fact $0 \in M(\lambda)$ for every positive dyadic rational λ

(ii) Since each M_i is symmetric so is each $M(\lambda)$. Thus

$$p(a) \leq \lambda$$

if and only if

$$a \in M(\lambda)$$

if and only if

$$-a \in -M(\lambda)$$

if and only if

$$-a \in M(\lambda)$$

if and only if

$$p(-a) \leq \lambda \ .$$

Since the dyadic rationals are dense, this is only possible if

$$p(a) = p(-a) \ .$$

(iii) It is sufficient, now, to show that

$$p(a + a') \leq p(a) + p(a') \ .$$

Now let $\varepsilon > 0$ be given.

Since the dyadic rationals are dense on the line, there are positive dyadic rationals λ and λ' such that

$$p(a) < \lambda < p(a) + \varepsilon/2$$

and

$$p(a') < \lambda' < p(a') + \varepsilon/2 \ .$$

It follows that

$$a \in M(\lambda)$$

and

$$a' \in M(\lambda')$$

so that

$$a + a' \in M(\lambda) + M(\lambda') \subset M(\lambda + \lambda')$$

so that

$$p(a+a') \leq \lambda + \lambda' \leq p(a) + p(a') + \varepsilon \quad .$$

Since $\varepsilon > 0$ is arbitrary, this is possible only if

$$p(a+a') \leq p(a) + p(a') \quad .$$

(iv) Let $a \in A$ and $\varepsilon > 0$ be given. Let λ be a dyadic rational such that

$$\lambda < \varepsilon \; .$$

Then $a + M(\lambda)$ is a neighborhood of 0 and if

$$a' \in a + M(\lambda) \; ,$$

we have,

$$a' - a \in M(\lambda)$$

or

$$p(a'-a) \leq \lambda < \varepsilon \; .$$

But

$$p(a) = p(a-a'+a') \leq p(a-a') + p(a')$$

gives

$$p(a) - p(a') \leq p(a-a') \quad .$$

Similarly,

$$p(a') - p(a) \leq p(a'-a) = p(a-a')$$

so that

$$|p(a') - p(a)| \leq p(a'-a) < \varepsilon \quad .$$

(v) This is by definition.

(5.8) This essentially completes the argument. For we have shown that for any neighborhood M of 0, there is a seminorm p such that

$$0 \in p^{-1}([0,1]) \subset M$$

while the continuity of p guarantees that $p^{-1}([0,1])$ is a neighborhood of 0.

(5.9) At this point we require a digression on the sums in the category of topological abelian groups. Let $\{A_\omega\}$ be a family of abelian groups. Then

$$A = \Sigma A_\omega$$

denotes, as usual, the subgroup of ΠA_ω consisting of all sequences with only a finite number of non-zero terms. If for each ω, p_ω is a seminorm on A_ω, let $p = (p_\omega)$ be the seminorm on A defined by

$$p(a_\omega) = \Sigma p_\omega a_\omega \; .$$

It is trivial to see that p is a seminorm on A and that if $u_\omega : A_\omega \longrightarrow A$ is the inclusion of the ωth summand, then

$$p_\omega = pu_\omega \quad .$$

Now supposing each A_i is a topological abelian group we endow A with the weak topology for the set of seminorms $p = (p_\omega)$ so defined as each p_ω runs independently over the set of seminorms on A_ω. We call this the <u>direct sum</u> topology on A, a term which we will justify later.

Before stating the next proposition, it will be convenient to say that of two seminorms p and q on A, p <u>refines</u> q if $p \geq q$. This is equivalent to the assertion that for all $\varepsilon > 0$ the cover of A by

$$\{M \subset A \mid a,a' \in M \Rightarrow p(a-a') < \varepsilon\}$$

refines the cover by

$$\{M \subset A \mid a,a' \in M \Rightarrow q(a-a') < \varepsilon\} \quad .$$

A <u>basis</u> of seminorms $\underline{p}$ is a set of seminorms with the property that for any seminorm q there is an $\varepsilon > 0$ and $p \in \underline{p}$ such that p refines εq.

(5.10) <u>Proposition</u>. Let A and B be topological groups and $f : |A| \longrightarrow |B|$ be a homomorphism of the underlying discrete groups. Then f is continuous iff for every continuous seminorm p on B, $p.f$ is a continuous seminorm on A.

Proof. Supposing f is continuous, so is $p.f$ while the seminorm property is evident. To go the other way, let $b \in B$. Every neighborhood of b is of the form $b + M$ where M is a neighborhood of 0. Let N be a neighborhood of 0 such that $N + N \subset M$ and let p be a seminorm on B such that

$$\{b' \mid p(b') < 1\} \subset N \quad .$$

Then if $p.f$ is a seminorm on A,

$$\{a \mid p(f(a)) < 1\} \subset f^{-1}(M)$$

is a neighborhood of 0. We have

$$a \in f^{-1}(b + M)$$

if and only if

$$f(a) \in b + M$$

if and only if

$$f(a) - b \in M$$

which is true if

$$p(f(a) - b) < 1 \quad .$$

Now if

$$\varepsilon = 1 - p(f(a) - b) > 0$$

and $a' \in A$ is such that

$$|pf(a) - pf(a')| < \varepsilon ,$$

we have

$$\begin{aligned} p(f(a') - b) &\leq p(f(a') - f(a)) + p(f(a) - b) \\ &< \varepsilon + 1 - \varepsilon = 1 \end{aligned}$$

so that

$$f(a') \in b + M \quad .$$

Thus

$$f^{-1}(b + M)$$

is an open set in A and hence f is continuous.

(5.11) <u>Proposition</u>. Let $\{A_\omega\}$ be a family of topological abelian groups. Then the topology on

$$A = \Sigma A_\omega$$

described above is the finest for which each inclusion

$$u_\omega : A_\omega \longrightarrow A$$

is continuous.

Proof. It is evident that if for each ω, p_ω is a continuous seminorm on A_ω and $p = (p_\omega)$, then $p_\omega = p.u_\omega$. Thus for each such p, $p.u_\omega$ is continuous and hence each u_ω is. Conversely, suppose q is any seminorm on A such that $q.u_\omega = p_\omega$ is continuous for each ω. Let $p = (p_\omega)$. Then for

$$a = (a_\omega) \in A$$

we have

$$a = \Sigma u_\omega(a)$$

so that

$$\begin{aligned} q(a) &= q(\Sigma\, u_\omega(a)) \\ &\le \Sigma\, qu_\omega(a) \\ &= \Sigma\, p_\omega(a) \\ &= p(a') \end{aligned}$$

and consequently p refines q. In particular, q is continuous in the topology defined by all such p.

Corollary. The direct sum topology is the categorical sum.

Proof. If B is a topological group and $f_\omega : A_\omega \longrightarrow B$ is a continuous homomorphism for all ω, let

$$f : \Sigma A_\omega \longrightarrow B$$

be defined in the usual way. Then $fu_\omega = f_\omega$. If q is a seminorm on B, qf is a seminorm on A, $qfu_\omega = qf_\omega$ is a seminorm on A_ω and hence qf is a seminorm on A with the direct sum topology. Thus f is a continuous homomorphism. The uniqueness of f is clear.

(5.12) Here and henceforth in this section, we let $T = \mathbb{R}/Z$. For a topological abelian group A, A^* is the group of continuous homomorphisms $A \longrightarrow T$ topologized in varying ways. Here we take the topology of compact convergence.

Proposition. Let $\{A_\omega\}$ be a family of topological abelian groups. Then there is a canonical map

$$\Sigma A_\omega{}^* \longrightarrow (\Pi A_\omega)^*$$

and that map is a natural isomorphism.

Proof. For each ω there is a projection

$$\pi_\omega : \Pi A_\omega \longrightarrow A_\omega$$

which dualizes to

$$A_\omega{}^* \longrightarrow (\Pi A_\omega)^* \quad .$$

The universal mapping property of the sum gives

$$\Sigma A_\omega{}^* \longrightarrow (\Pi A_\omega)^*$$

which is evidently canonical and natural. Now suppose

$$f : \Pi A_\omega \longrightarrow T$$

is a continuous homomorphism. Let

$$u_\psi : A_\psi \longrightarrow \Pi A_\omega$$

be the map which is the identity on the ψ coordinate and 0 on all others. Let

$$f_\omega = f . u_\omega \quad .$$

I first claim that only finitely many f_ω are different from 0. For let M be the neighborhood of 0 in T represented by the interval $\left(-\frac{1}{4}, \frac{1}{4}\right)$. Evidently M contains no non-zero subgroup of T. [Repeated doubling of a $t \in (0, \frac{1}{4})$ will eventually, by the archimedean property of the reals, put it into the interval $\left(\frac{1}{4}, \frac{1}{2}\right)$ Similarly for a $t \in \left(\frac{1}{4}, 0\right)$.] Thus the image of a non-zero subgroup cannot lie in M. Now $f^{-1}(M)$ is to be a neighborhood of 0 in ΠA_ω. This means that

$$f^{-1}(M) \supset \Pi M_\omega$$

where M_ω is a neighborhood of 0 in M_ω and except for finitely many ω, say $\omega = \omega_1, \ldots, \omega_n$ $M_\omega = A_\omega$. In particular

$$f^{-1}(M) \supset B = \pi B_\omega$$

where

$$B_\omega = \begin{cases} 0, & \omega = \omega_1, \ldots, \omega_n \\ A_\omega, & \text{otherwise} . \end{cases}$$

Since B is a subgroup and

$$f(B) \subset M$$

it follows that $f(B) = 0$. In particular since

$$u_\omega(A_\omega) \subset B$$

for

$$\omega \neq \omega_1, \ldots, \omega_n .$$

it follows that

$$f_\omega = 0 ; \quad \omega \neq \omega_1, \ldots, \omega_n \quad .$$

Then if

$$a = (a_\omega) \in A \quad ,$$

the difference

$$a' = a - \sum\{u_\omega a_\omega \mid \omega = \omega_1, \ldots, \omega_n\}$$

belongs to B so that $f(a') - 0$. Hence

$$\begin{aligned} fa &= f(\Sigma u_\omega a_\omega) \\ &= \Sigma f u_\omega a_\omega \\ &= \Sigma f_\omega a_\omega \\ &= \Sigma f_\omega \pi_\omega a \end{aligned}$$

so that

$$f = \Sigma f_\omega \pi_\omega \quad .$$

This shows that the canonical map is at least an algebraic isomorphism.

Now let M_i be the neighborhood of 0 in T represented by the interval

$$(-2^{-i}, 2^{-i})$$

for $i \geq 2$. Then the family of all M_i is a neighborhood base at 0. In particular

$$M_2 = M$$

(as defined above). Now a neighborhood base at 0 in A^* consists of

$$\{f \mid f(X) \subset M_i\}$$

as X ranges over compact sets and i over all integers ≥ 2. Since $X \cup \{0\}$ is compact when X is and

$$\{f \mid f(X) \subset M_i\} \supset \Big\{f \mid f(X \cup \{0\}) \subset M_i\Big\}$$

we can restrict to X which contain 0. If

$$f(2^{i-2}X) \subset M_2 = M \quad ,$$

then, since for all $x \in X$, and integer i,

$$0 \leq j \leq 2^{i-2} \, ,$$

$$jx = jx + (2^{i-2} - j)0 \in 2^{i-2}X$$

then for all such j,

$$f(jx) \in M \quad .$$

If $f(x) \not\subset M_i$, there is same $j < 2^{i-2}$ such that

$$f(jx) \not\subset M$$

which is a contradiction. In fact, j may taken as 2 raised to the power

$$4 + [\log_2 |r|]$$

where r is the absolutely least residue of x (mod 1).
Thus

$$\{f \mid f(X) \subset M_i\} \supset \{f \mid f(2^{i-2}X) \subset M\} \quad .$$

Since $2^{i-2}X$ is compact when X is — it is the image of the 2^{i-2}nd power of X under the addition map of that power of A to A — the sets of the form

$$\{f \mid f(X) \subset M\}$$

as X ranges over the compact sets in A form neighborhood base at 0 in A^*.
Now let p be the seminorm on T which assigns to x, 4 times its absolutely least least residue, modulo 1. Then

$$x \in M \iff p(x) < 1 \quad .$$

Thus

$$f(X) \subset M$$

if and only if $pf < 1$ on all of X. If we let $\hat{X}$ denote the seminorm defined by

$$\hat{X}(f) = \sup\{pf(X) \mid x \in X\}$$

then f belongs to

$$\{g \mid g(X) \subset M\}$$

if and only if

$$\hat{X}(f) < 1 \quad .$$

Thus the seminorms $\hat{X}$ determine, as X ranges over the compact sets of A, a basis of seminorms of A^*.

Now returning to

$$A = \Pi A_\omega \quad ,$$

if $X \subset A$ is compact, $\pi_\omega X$ is compact in A_ω. Evidently

$$X \supset X^\# = \Pi \pi_\omega X$$

so that

$$\{f \mid f(X) \subset M\} \supset \{f \mid f(X^\#) \subset M\}$$

or, equivalently, $\hat{X}$ is refined by $X^{\#\wedge}$.

Now if

$$Y = X \cup \{0\} \cup (-X)$$

Y is also compact and $\hat{Y}$ refines $\hat{X}$. Thus a basis of seminorms of $A = \Pi A_\omega$ consists of $\hat{X}$ where X is compact

$$X = \Pi \pi_\omega X \quad ,$$

$$0 \in X \quad ,$$

and

$$X = -X \quad .$$

I claim that under these hypotheses

$$\hat{X} = ((\pi_\omega X)^\wedge) \quad .$$

(5.13) At this point, we require,

Lemma. Let $n > 1$ and $x_1, \ldots, x_n \in T$ be such that the absolutely least residue of x_i is positive,

$$\sum_{i=1}^{j} px_i < 1 \; , \qquad j = 1, \ldots, n-1 \quad ,$$

and

$$px_n < 1 \; .$$

Then the absolutely least residue of

$$\sum_{i=1}^{n} x_i$$

is positive. Moreover,

$$\sum_{i=1}^{n} px_i = p\left(\sum_{i=1}^{n} x_i\right) \; .$$

Proof. We will confuse x_i and its absolutely least residue. When $n = 2$, the hypotheses are that

$$0 \leq x_1 < \frac{1}{4}$$

$$0 \leq x_2 < \frac{1}{4}$$

and, of course,

$$x_1 + x_2 < \frac{1}{2} \quad .$$

This implies that $x_1 + x_2$ is its own absolutely least residue and hence that that residue is positive and moreover that

$$\begin{aligned} p(x_1 + x_2) &= 4(x_1 + x_2) \\ &= 4x_1 + 4x_2 \\ &= p(x_1) + p(x_2) \quad . \end{aligned}$$

For larger n, we may suppose inductively that the assertion is valid for $n-1$. Then the hypotheses are satisfied by

$$\sum_{i=1}^{n-1} x_i \quad \text{and} \quad x_n \; .$$

Hence the absolutely least residue of $\sum_{i=1}^{n} n_i$ is positive and

$$\begin{aligned} p\left(\sum_{i=1}^{n} x_i\right) &= p\left(\sum_{i=1}^{n-1} x_i\right) + px_n \\ &= \sum_{i=1}^{n-1} px_i + px_n \\ &= \sum_{i=1}^{m} px_i \end{aligned}$$

(5.14) Now we return to the proof of 5.12. We suppose that is a compact subset of A such that

$$X = \Pi\pi_\omega X \; ;$$

$$0 \in X$$

and

$$X = -X \; .$$

Let

$$f : A \longrightarrow T$$

be such that

$$\hat{X}(f) \geq 1 \; .$$

Then is there some $x \in X$ such that

$$p f(x) \geq 1 \; .$$

Now f is a finite sum

$$f(x) = \Sigma f u_\omega \pi_\omega x$$

and

$$\begin{aligned} 1 \leq p f(x) &= p(\Sigma f u_\omega \pi_\omega n) \\ &\leq \Sigma p f u_\omega \pi_\omega x \\ &\leq \Sigma \sup \{p f u_\omega n_\omega | x_\omega \in X_\omega\} \\ &= \Sigma \hat{X}_\omega (f u_\omega) \quad . \end{aligned}$$

Thus

$$\{f \mid \hat{X}(f) < 1\} \supset \{f \mid \Sigma \hat{X}_\omega(f U_\omega) < 1\} .$$

To go the other way, suppose

$$\Sigma \hat{X}_\omega(f u_\omega) \geq 1 .$$

Then there are $x_\omega \in X_\omega$ such that

$$\Sigma\, p f u_\omega x_\omega \geq 1 .$$

(of course X and the X_ω are compact, so these sups are actually attained. The arguments would complicate somewhat but the same principle would work even if they weren't.) Now since X and hence X_ω is symmetric, we can suppose that the absolutely least residue of each $f u_\omega x_\omega$ is positive. Since $f u_\omega = 0$ for all but finitely many ω, the sum is finite. Suppose the set of indices involved is $\omega_1, \ldots, \omega_n$. Now we may suppose, by induction, that

$$\Sigma \{p f u_\omega n_\omega \mid \omega = \omega_1, \ldots, \omega_{n-1}\} < 1 .$$

By our lemma, this means either that

$$p f u_{\omega_n} x_{\omega_n} \geq 1$$

or that

$$\begin{aligned} & p\Big(\Sigma \{f u_\omega x_\omega \mid \omega = \omega_1, \ldots, \omega_n\}\Big) \\ = {} & \Sigma \{p f u_\omega x_\omega \mid \omega = \omega_1, \ldots, \omega_n\} \\ \geq {} & 1 . \end{aligned}$$

In the first case, let $x \in X$ have 0 in every coordinate but the ω_n^{th} and x_{ω_n} there. In the second case let x have x_{ω_i} in the ω_i^{th} coordinate and 0 elsewhere. In either case $x \in X$ while

$$f(x) > 1$$

so that

$$\hat{X}(f) > 1 .$$

This shows that

$$\{f \mid \hat{X}(f) < 1\} = \{f \mid \Sigma \hat{X}_\omega(f u_\omega) < 1\} .$$

Thus the topologies on ΣA_ω^* and $(\Pi A_\omega)^*$ are identical and these groups are isomorphic.

(5.15) <u>Proposition</u>. A direct sum of complete groups is complete.

Proof. Let $\{A_\omega\}$ be a family of complete groups and

$$A = \Pi A_\omega$$

and A' be the same abelian group but topologized by the box topology — the one in which the product of any family of open sets is open. First I claim that A' is a topological group. In fact a neighborhood of (a_ω) contains a set of the form

$$\Pi(a_\omega + M_\omega) = (a_\omega) + \Pi M_\omega$$

where for ω, M_ω is a neighborhood of 0 in A_ω. Moreover if N_ω is a neighborhood of 0 such that

$$N_\omega - N_\omega \subset M_\omega ,$$

$$\Pi N_\omega - \Pi N_\omega \subset \Pi M_\omega .$$

Next I claim that A' is complete. In fact, by an obvious additive analogue of I.2.5, it is sufficient, since A is complete, to show that there is a neighborhood base at 0 in A' consisting of sets closed in A. But of course the M_ω may be taken as closed neighborhoods of 0 in the A_ω, whence the ΠM_ω is closed in A. Now let B be the subgroup (of A') whose elements are those of the direct sum. I claim that B is a closed subgroup. For if

$$a = (a_\omega) \notin B ,$$

then $a_\omega \neq 0$ for infinitely many ω. For all these ω let M_ω be a symmetric neighborhood of 0 such that

$$a_\omega \notin M_\omega .$$

Let M_ω be an arbitrary neighborhood of 0 for all other coordinates and let

$$M = \Pi M_\omega .$$

Then

$$(a_\omega) + M$$

is a neighborhood of (a_ω). If

$$(b_\omega) \in (a_\omega) + M$$

then when

$$a_\omega \neq 0 ,$$

we have

$$a_\omega \notin M$$

so that

$$-a_\omega \notin M$$

and then

$$0 \notin a_\omega + M$$

so that

$$b_\omega \neq 0 .$$

Thus

$$(b_\omega) \notin B .$$

This means

$$(a_\omega) + M$$

is disjoint from B. Hence the complement of B is open in A' and B is closed. In particular B is also complete.

The topology on B can be described as the one generated by the seminorms

$$p(a_\omega) = \sup (p_\omega a_\omega)$$

where for each ω, p_ω is a seminorm on A_ω. (Note that the sup is finite.) For if

$$M = B \cap \Pi M_\omega$$

is a neighborhood of 0 in B where M_ω is a neighborhood of zero in A_ω, let p_ω be a seminorm on A_ω such that

$$M_\omega \subset \{a_\omega | pa_\omega < 1\} .$$

Then

$$(a_\omega) \in \Pi M_\omega$$

if and only if

$$p_\omega a_\omega < 1$$

for all ω, if and only if

$$\sup p_\omega a_\omega < 1 .$$

Thus the topology on B is coarser than that of the direct sum (which, recall, is determined by Σp_ω where each p_ω is a seminorm on A_ω). Now we may infer the completeness of the direct sum from another application of I.2.5. It is the sufficient to show that the basic neighborhood

$$M = \{(a_\omega) | \Sigma p_\omega a_\omega \leq 1\}$$

is closed in B. Suppose

$$(a_\omega) \notin M$$

which means that

$$\Sigma p_\omega a_\omega > 1 .$$

Choose $\varepsilon > 0$ so that

$$\Sigma p_\omega a_\omega > 1 + \varepsilon .$$

Since $(a_\omega) \in \Sigma A_\omega$, this sum is actually finite. Let $\omega_1, \ldots, \omega_n$ be the finite number of indices involved. Then for $i = 1, \ldots, n$, define seminorms q_{ω_i} by

$$q_{\omega_i} = np_{\omega_i}/\varepsilon .$$

Let

$$q_\omega = 0$$

for indices $\omega \neq \omega_1, \ldots, \omega_n$. Then evidently q_ω is a continuous seminorm on A_ω. Moreover, if

$$(b_\omega) \in B$$

is such that

$$\sup q_\omega b_\omega < 1 ,$$

we have

$$q_{\omega_i} b_{\omega_i} = np_{\omega_i} b_{\omega_i} \varepsilon < 1$$

or

$$p_{\omega_i} b_{\omega_i} < \varepsilon/n$$

for $i - 1, \ldots, n$. Then we have,

$$\sum p_\omega(a_\omega + b_\omega) \geq \sum_{i=1}^{n} p_{\omega_i}\left(a_{\omega_i} + b_{\omega_i}\right)$$

$$> \sum_{i=1}^{n} p_{\omega_i} a_{\omega_i} - \sum_{i=1}^{n} p_{\omega_i} b_{\omega_i}$$

$$> 1 + \varepsilon - \sum_{i=1}^{n} \varepsilon/n$$

$$= 1 + \varepsilon - \varepsilon$$
$$= 1 .$$

If

$$N = \{b_\omega | \sup q_\omega b_\omega < 1\} ,$$

we have that

$$((a_\omega) + N) \cap M = \emptyset .$$

and M is closed. This completes the proof.

(5.16) <u>Proposition</u>. Let

$$A = \Sigma A_\omega .$$

Then every compact set in A lies in a sum of finitely many factors.

Proof. Let X be a compact subset of A. Let ψ be an index such that there is an element $(a_\omega) \in X$ with

$$a_\psi \neq 0 .$$

Let M_ψ be a neighborhood of 0 with

$$a_\psi \notin M_\psi$$

and let p_ψ be a seminorm on A_ψ such that

$$M_\psi \supset \{a \in A_\psi | p_\psi(a) < 1\}$$

whence

$$p_\psi(a_\psi) \geq 1 .$$

Now assuming this to be possible for at least countably many coordinates, say

$$\psi = \psi_1, \psi_2, \ldots ,$$

then we can find such seminorms

$$p_{\psi_1}, p_{\psi_2}, \ldots .$$

Let q be the seminorm on A which has the seminorm

$$q_i = i p_{\psi_i}$$

in the ith coordinate and 0 elsewhere. By hypothesis there is for each $i = 1,2,\ldots$ an element

$$(a_{\omega i}) \in X$$

such that

$$p_{\psi_i} a_{\psi_i} i \geq 1$$

so that

$$q_i a_{\psi_i} i \geq i$$

so that

$$q(a_{\omega i}) = \sum q_j a_{\psi_j} j$$
$$\geq q_i a_{\psi_i} i$$
$$\geq i .$$

But then q is a continuous unbounded real valued map on the compact set X which is impossible.

(5.17) Proposition. Let A and B be topological abelian groups. Then the canonical bijection

$$A \oplus B \longrightarrow A \times B$$

is an isomorphism.

Proof. This map exists in any pointed category and is, of course, a bijection of the underlying abelian groups. If p and q are seminorms on A and B, respectively,

$$\{(a,b) \mid pa + qb < 1\} \supset \{(a,b) \mid pa < \tfrac{1}{2}\} \cap \{(a,b) \mid qb < \tfrac{1}{2}\} .$$

The left hand side is a basic neighborhood of 0 in $A \oplus B$ while the right hand side is the intersection of two such in $A \times B$. Thus the map is also open.

Corollary 1. The canonical map

$$(A \oplus B)^* \longrightarrow A^* \times B^*$$

is an isomorphism.

Proof. By the canonical map is meant the one whose A^* coordinate, for example, is gotten by dualizing the injection

$$A \longrightarrow A \oplus B .$$

At any rate, we have

$$\begin{aligned} (A \oplus B)^* &\cong (A \times B)^* \\ &\cong A^* \oplus B^* \\ &\cong A^* \times B^* . \end{aligned}$$

Corollary 2. If $A_1, \ldots, A_n$ are topological abelian groups, the canonical map

$$(\Sigma A_i)^* \longrightarrow \Pi A_i^*$$

is an isomorphism.

(5.18) Proposition. Let $\{A_\omega\}$ be a family of topological abelian groups. Then the canonical map

$$(\Sigma A_\omega)^* \longrightarrow \Pi A_\omega^*$$

is an isomorphism.

Proof. The fact that the canonical map is a continuous bijection is trivial. A basic neighborhood of 0 in

$$(\Sigma A_\omega)^*$$

is

$$N(X,M) = \{f \mid f(X) \subset M\}$$

where

$$X \subset \Sigma A_\omega$$

is compact and M is a neighborhood of 0 in T. It follows from 5.16 that X is contained in a finite sum of A_ω, say

$$X \subset A_{\omega_1} \oplus \ldots \oplus A_{\omega_n} = A(\omega_1, \ldots, \omega_n) .$$

Thus N(X,M) is the inverse image of the set of the same name in

$$A(\omega_1, \ldots, \omega_n)^*$$

under the dual of the canonical inclusion

$$A(\omega_1, \ldots, \omega_n) \longrightarrow \Sigma A_\omega \quad .$$

From the commutativity of

$$\begin{array}{ccc} (\Sigma A_\omega)^* & \longrightarrow & \Pi A_\omega^* \\ \downarrow & & \downarrow \\ A(\omega_1, \ldots, \omega_n)^* & \longrightarrow & A_{\omega_1}^* \times \ldots \times A_{\omega_n}^* \end{array}$$

together with the previously established fact that the lower arrow is an isomorphism, it follows that the inverse image of that set in ΠA_ω^* is open. Since the upper arrow is a bijection, it follows that inverse image is the image of $N(X,M)$ under the canonical map.

(5.19) Before continuing, it is necessary to discuss Pontijagin duality for compact and discrete groups. If D is a discrete group, the group

$$D^* = (D,T)$$

is topologized by the product topology, i.e. as a subspace of T^D.

<u>Proposition</u>. The group of homomorphisms of $D \longrightarrow T$ is closed in T^D and hence compact.

Proof. If $x,y \in D$, the map

$$T^D \longrightarrow T$$

defined by

$$f \longmapsto f(x) + f(y) - f(x+y)$$

factors

$$T^D \longrightarrow T \times T \times T \longrightarrow T$$

where the first map is projected onto the factors corresponding to x,y and $x+y$ and the second takes

$$(t_1,t_2,t_3) \longmapsto t_1 + t_2 - t_3 \quad .$$

Both are continuous and so the inverse of 0,

$$\{f \mid f(x) + f(y) = f(x+y)\} \quad ,$$

is closed. The intersection of these over all $x,y \in D$ is then closed.

(5.20) If C is a compact group, we let

$$C^* = (C,T)$$

with the discrete topology. We use here one highly non-trivial fact from the theory of topological groups.

<u>Theorem</u>. If C is compact, then for any $x \in C$, $x \neq 0$, there is a continuous homomorphism

$$\phi : C \longrightarrow T$$

such that $\phi(x) \neq 0$.

Proof. See [Hewitt-Ross], 22.17 on p.345.

(5.21) This implies that for any compact group C, the canonical map

$$C \longrightarrow C^{**}$$

is injective. The analogous fact for discrete groups follows readily from the well known fact that T is an injective cogenerator in the category of abelian groups.

The continuity, for a compact C, of

$$C \longrightarrow C^{**}$$

follows immediately from the pointwise convergence topology of the latter and the fact that every map in C^* is continuous. The analogous fact for discrete groups requires, of course, no proof.

(5.22) Proposition.

$$\mathbb{Z}^* \cong T \; ; \quad T^* \cong \mathbb{Z} \; .$$

Proof. The first is obvious. As for the second, a map $\phi : T \longrightarrow T$ is equivalent to a map $\psi : \mathbb{R} \longrightarrow T$ such that $\phi(\mathbb{Z}) = 0$. The kernel K of such a map either contains a smallest positive number λ or else contains arbitrary small positive numbers. In the latter case, let $x \in \mathbb{R}$ and $\varepsilon > 0$. Then there is a $\lambda \in K$ with $0 < \lambda < \varepsilon$. Then if n is chosen so that $n\lambda \le x \le (n+1)\lambda$, $|n\lambda - x| < \varepsilon$ and $n\lambda \in K$. Thus K is dense and since T is separated, $K = \mathbb{R}$ and $\psi = 0$ whence $\phi = 0$. Otherwise, there is a smallest $\lambda \in K$ Choose m such that

$$n\lambda \le 1 < (n+1)\lambda \; .$$

Then

$$0 \le n\lambda - 1 < \lambda \; .$$

Since $1 \in K$ and $n\lambda \in K$ this is only possible if $n\lambda = 1$ or

$$\lambda = \frac{1}{n} \; .$$

Now choose $\varepsilon > 0$ so that

$$|x| < \varepsilon \Longrightarrow \psi(x) \in \left(-\frac{1}{4}, \frac{1}{4}\right) .$$

Choose an integer $k > 1$ such that

$$|2^{-k}\lambda| < \varepsilon$$

so that for $j \ge k$,

$$\psi(2^{-j}\lambda) \in \left(-\frac{1}{4}, \frac{1}{4}\right) .$$

Now let a be the absolutely least residue of $\psi(2^{-k}\lambda)$. Then the absolutely least residue of $\psi(2^{-k-1}\lambda)$ can only be either $\frac{a}{2}$ or $\frac{a}{2} \pm \frac{1}{2}$ (depending on whether $a < 0$ or $a > 0$). But if $a \in (0, \frac{1}{4})$, $a - \frac{1}{2} < -\frac{1}{4}$ and similarly in the other case. Thus the absolutely least residue of $\psi(2^{-k-1}\lambda)$ can only be $\frac{a}{2}$. Similarly, the absolutely least residue of $\psi(2^{-k-j}\lambda)$ is

$$2^{-j}a = n2^k a \, 2^{-j-k}\lambda \; .$$

That is

$$x = 2^{-k-j}\lambda$$

implies

$$\psi(x) = n2^k ax .$$

This continues to be true when x is a linear combination of such elements and such x are clearly dense in T . Hence

$$\psi(x) = n2^k ax$$

for all $x \in R$. Moreover,

$$0 = \psi(\lambda) = \psi(2^k 2^{-k}\lambda) = 2^k a$$

so that $2^k a$ is an integer. If

$$|2^k a| > 1 ,$$

$$\psi\left(\frac{\lambda}{|2^k a|}\right) \equiv 0 \pmod 1$$

so that $\frac{\lambda}{|2^k a|}$ is a smaller positive element in the kernel. Hence ψ is either multiplication by n or by $-n$. This shows that the canonical map

$$\mathbb{Z} \longrightarrow (T, T)$$

is surjective. That it is injective is clear and hence it is an isomorphism.

(5.23) It is clear from (5.12) and (5.18) that sums and products of copies of $\mathbb{Z}$ and T are also reflexive.

Proposition. A compact group is isomorphic to a closed subgroup of a power of T .

Proof. Let C be compact. The natural map

$$C \longrightarrow T^{C^*}$$

is injective. Since C is compact, it is homeomorphic, hence isomorphic, to its image.

(5.24) Proposition. Let T be embedded in the compact group C . Then T is a direct summand of C .

Proof. It is clearly sufficient to show that the map

$$C^* \longrightarrow T^* ,$$

induced by the embedding, is surjective. Suppose, to the contrary, that the proper subgroup $D \subset T^*$ is the image. Since T is a cogenerator, there is a non-zero character on T^*/D . This means there is a non-zero character on T^* which vanishes on D or that

$$T^{**} \longrightarrow D^*$$

is not injective. But we have a commutative diagram

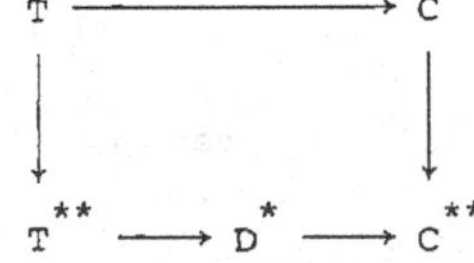

in which the top and right hand map are injections and the left an isomorphism from which it follows that the bottom map is an injection as well.

(5.25) Proposition. The group T is injective in the category of compact groups.

Proof. The proof is standard. If

$$C_1 \longrightarrow C_2$$

is an injection and $\phi : C_1 \longrightarrow T$ is a map, form the pushout

$$\begin{array}{ccc} C_1 & \longrightarrow & C_2 \\ \downarrow & & \downarrow \\ T & \longrightarrow & C \end{array}$$

This is the compact group $T \times C_2$ modulo the compact, hence closed subgroup, consisting of all

$$\{ (-\phi(x), x) \mid x \in C_1 \} .$$

In other words, the pushout is the same as of the underlying abelian groups. Thus T is a subgroup of C and the preceding proposition completes the argument.

(5.26) <u>Proposition</u>. Every compact group is reflexive.

Proof. If C is compact, embed C in a power of T, say T^n (n needn't be finite). Then

$$0 \longrightarrow C \longrightarrow T^n \longrightarrow T^n/C \longrightarrow 0$$

is exact and T^n/C is compact. Then since T is injective in both categories of compact and discrete abelian groups we have a commutative diagram with exact rows,

$$\begin{array}{ccccccccc} 0 & \longrightarrow & C & \longrightarrow & T^n & \longrightarrow & T^n/C & \longrightarrow & 0 \\ & & \downarrow & & \downarrow & & \downarrow & & \\ 0 & \longrightarrow & C^{**} & \longrightarrow & (T^n)^{**} & \longrightarrow & (T^n/C)^{**} & \longrightarrow & 0 \end{array}$$

with the vertical maps injections. It follows from (5.12) and (5.18) that the middle one is an isomorphism. Since the bottom composite is 0, it is an easy diagram chase to work out that the left hand map must be surjective. Since C and C^{**} are compact, it is an isomorphism.

(5.27) <u>Proposition</u>. Every discrete group is reflexive.

Proof. If D is discrete, D^* is compact. The canonical map

$$D^* \longrightarrow D^{***}$$

is, by the preceding, an isomorphism. Since it is always true (in any closed category) that the composite of that map with the dual of the natural map

$$D \longrightarrow D^{**}$$

is the identity on D^*, it follows that the dual of that map is an isomorphism. Now since T is a cogenerator, this map is injection. Thus there is an exact sequence

$$0 \longrightarrow D \longrightarrow D^{**} \longrightarrow D^{**}/D \longrightarrow 0 .$$

Then we have a commutative diagram with exact rows,

$$\begin{array}{ccccccccc} 0 & \longrightarrow & D & \longrightarrow & D^{**} & \longrightarrow & D^{**}/D & \longrightarrow & 0 \\ & & \downarrow & & \downarrow & & \downarrow & & \\ 0 & \longrightarrow & D^{**} & \longrightarrow & D^{****} & \longrightarrow & (D^{**}/D)^{**} & \longrightarrow & 0 \end{array}$$

in which the vertical maps are injections, the middle an isomorphism and hence so is the left hand one.

This shows that the duality is valid between the compact and the discrete groups.

(5.28) Proposition. The dual of the group $\mathbb{R}$ of real numbers, equipped with the topology of uniform convergence on compact sets, is $\mathbb{R}$.

Proof. If

$$\phi : \mathbb{R} \longrightarrow T$$

is a map, let

$$\lambda = \inf \{x \mid x > 0 \quad \& \quad \phi(x) = 0\} .$$

By continuity, $\phi(\lambda) = 0$ and $\lambda = 0$ only when $\phi = 0$. Otherwise let

$$\psi(1/\lambda) : \mathbb{R} \longrightarrow T$$

by

$$\psi(1/\lambda)(x) = x/\lambda \pmod 1 .$$

We have an exact sequence

$$0 \longrightarrow \lambda\mathbb{Z} \longrightarrow \mathbb{R} \longrightarrow \mathbb{R}/\lambda\mathbb{Z} \cong T \longrightarrow 0$$

and $\phi(\lambda\mathbb{Z}) = 0$, ϕ induces a map

$$\tilde{\phi} : T \longrightarrow T .$$

We know from 5.22 that $\tilde{\phi}$ is multiplication by an integer. Since $\ker \phi = \ker \psi$, $\tilde{\phi}$ is an injection. The only maps $T \longrightarrow T$ which are injections are the identity and the inverse. Thus

$$\phi = \pm \psi(1/\lambda) .$$

If we define, for all $\lambda \in \mathbb{R}$

$$\psi(\lambda) : \mathbb{R} \longrightarrow T$$

by

$$\psi(\lambda)(x) = \lambda x \pmod 1$$

then the above shows that every ϕ is of the form $\psi(\lambda)$ for a$\lambda \in \mathbb{R}$ which is clearly unique.

A basis for the compact sets in $\mathbb{R}$ consists of the intervals $[-n, n]$, n a natural number. Evidently $\psi(\lambda)$ takes that set into $\left(-\frac{1}{4}, \frac{1}{4}\right)$ iff $|\lambda| < 1/4n$ so that the basic neighborhoods of 0 are just the usual ones.

(5.29) We are now ready to apply the theory.

We consider six passibilities for the subcategory $\underline{C}$ and as many for $\underline{D}$. Each leads to a different choice of $\underline{A}$, of duality and finally, of $\underline{G}$. We let $\underline{C}$ consist of all groups of the form

$$C \times \mathbb{R}^n \times \mathbb{Z}^m$$

and $\underline{D}$ consist of all groups of the form

$$D \oplus n.\mathbb{R} \oplus m.T$$

subject to

i) C is compact and D discrete;

ii) $n = 0$, n is finite, or n is at most countable;

iii) m is finite, or m is at most countable.

The three choices in (ii) and two in (iii) give the six possibilities.

A compact group can be embedded, as mentioned above, in a power of T. A power of $\mathbb{R}$ can of course be embedded in a power of $\mathbb{R}$. A power of $\mathbb{Z}$ can be embedded in a product of discrete groups. Thus for all possibile choices of $\underline{C}$, every group in $\underline{C}$ can be embedded in a product of groups in $\underline{D}$. Whatever choice of $\underline{C}$ and $\underline{D}$ is made, we let $\underline{A}$, as usual, consist of all groups that can be embedded in a product of groups in $\underline{D}$.

(5.30) We define the functor

$$(-,-) : \underline{C}^{op} \times \underline{D} \longrightarrow \underline{D}$$

by letting

$$(C \times \mathbb{R}^m \times \mathbb{Z}^n , \quad D \oplus p.\mathbb{R} \oplus q.T)$$

be the direct sum of nine terms as follows.

i) (C,D) is the group of continuous maps $C \longrightarrow D$ topologized discretely;

ii) $(C, p.\mathbb{R}) = 0$;

iii) $(C, q.T) = q.C^*$;

iv) $(\mathbb{R}^m, D) = 0$;

v) $(\mathbb{R}^m, n.\mathbb{R}) = (m \times n).\mathbb{R}$;

vi) $(\mathbb{R}^m, q.T) = (m\times q).\mathbb{R}$;

vii) $(\mathbb{Z}^n, D) = n.D$;

viii) $(\mathbb{Z}^n, p.\mathbb{R}) = (m \times p).\mathbb{R}$;

ix) $(\mathbb{Z}^n, q.T) = (n \times q).T$.

We wish to establish that these are the correct underlying groups. For example, in (iii) the image of a map $C \longrightarrow q.T$ lies in a compact subgroup of T . By (5.16) it lies in $q_o.T$ for same finite subset $q_o \subset q$. Hence

$$\begin{aligned} \mathrm{Hom}(C,q.T) &\cong \underrightarrow{\lim}\, \mathrm{Hom}(C,q_o.T) \\ &\cong \underrightarrow{\lim}\, \mathrm{Hom}(C,T^{q_o}) \\ &\cong \underrightarrow{\lim}\, \mathrm{Hom}(C,T)^{q_o} \\ &\cong \underrightarrow{\lim}\, q_o.\mathrm{Hom}(C,T) \\ &\cong q.\mathrm{Hom}(C,T) . \end{aligned}$$

Here Hom $(-, -)$ refers to the abelian group valued hom. Both groups $\mathbb{R}^m$ and $\mathbb{Z}^n$ are generated by compact subsets, $[-1, 1]^m$ and $\{-1, 0, 1\}^n$ respectively. Thus any map from either of these groups to a direct sum must also factor through a finite sum. Analogously (and, if fact, dually) the groups in $\underline{D}$ all have a neighborhood M of 0 that contains no non-zero subgroup. If f is a map to such a group, any subgroup contained in $f^{-1}(M)$ lies in the kernel of f . Thus a continuous map from a power $\mathbb{R}^m$ or $\mathbb{Z}^n$ must contain in its kernel the product of all but finitely many. Then a dual argument to the above shows that in each case, the discrete abelian group underlying (A,B) is $\mathrm{Hom}(A,B)$.

(5.31) <u>Proposition</u>. Let $C \in \underline{C}$ and $D \in \underline{D}$. Then (C,D) is $\mathrm{Hom}(C,D)$ topologized by uniform convergence on compact subsets (compact convergence).

Proof. We have already established that the abelian group underlying (C,D) is $\mathrm{Hom}(C,D)$ We consider the cases separately.

i) When C is compact and D discrete,

$$\{ f : C \longrightarrow D \mid f(C) = 0 \} = 0$$

is open in the compact convergence topology so that the topology is discrete.

ii) There is nothing to prove.

iii) The group $q \cdot C^*$ is discrete. On the other hand, $q \cdot T$ has a neighborhood M of 0 without any small subgroups. For example, if for all $\alpha \in q$, $M_\alpha = (-1/4, 1/4)$,

$$M = \Gamma(M_\alpha)$$

will do. Thus $\mathrm{Hom}(C, q \cdot T)$ is discrete in the compact open topology.

iv) Again, there is nothing to prove.

v) We must show that the abstract isomorphism

$$(n \times m) \cdot \mathbb{R} \longrightarrow (\mathbb{R}^n, m \cdot \mathbb{R})$$

is a homeomorphism.

Here n and m may be finite or countable. We consider the latter case as the other one is easier. We first show it is open. A neighborhood of 0 on the left is of the form

$$\Gamma(-r_{\alpha\beta}, r_{\alpha\beta})$$

where $\{r_{\alpha\beta}\}$ is a doubly indexed sequence of positive real numbers. The neighborhood so determined consists of all doubly indexed sequences

$$f = \{f_{\alpha\beta}\}$$

of real numbers such that

i) Only finitely many $f_{\alpha\beta} \neq 0$, and

ii) $\sum_{\alpha,\beta} |f_{\alpha\beta}/r_{\alpha\beta}| < 1$.

Given such a sequence, let

$$s_\alpha = 2^\alpha \sup\{1, 1/r_{\beta\gamma} \mid \beta, \gamma \le \alpha\}$$

and

$$t_\alpha = 2^{-\alpha} \inf\{1, r_{\beta\gamma} \mid \beta, \gamma \le \alpha\} .$$

Then for $\alpha \le \beta$, we have

$$s_\alpha \ge 2^\alpha$$

and

$$t \le 2^{-\beta} r_{\alpha\beta}$$

while $\alpha \ge \beta$ implies

$$s_\alpha \ge 2^\alpha / r_{\alpha\beta}$$

and

$$t_\beta \le 2^{-\beta}$$

and in either case

$$t_\beta / s_\alpha \le 2^{-\alpha-\beta} r_{\alpha\beta} .$$

On the right hand side the set

$$\left\{ f \mid f\left(\Pi\,[-s_\alpha, s_\alpha]\right) \subset \Gamma(-t_\beta, t_\beta) \right\}$$

is a neighborhood of 0 . Every

$$f = (f_{\alpha\beta})$$

in that set must in particular have the property that

$$f_{\alpha\beta}[-s_\alpha, s_\alpha] \subset (-t_\beta, t_\beta)$$

or

$$|f_{\alpha\beta} s_\alpha| < t_\beta$$

$$|f_{\alpha\beta}| < t_\beta/s_\alpha \le 2^{-\alpha-\beta} r_{\alpha\beta}$$

so that

$$|\Sigma f_{\alpha\beta}/r_{\alpha\beta}| \le \Sigma|f_{\alpha\beta}/r_{\alpha\beta}| \le 1 .$$

This shows that the map is open. To go the other way, every compact set in $\mathbb{R}^n$ is contained in one of the form

$$\Pi [-s_\alpha, s_\alpha]$$

while sets of the form

$$\Gamma(-t_\beta, t_\beta)$$

are a basis for the open sets in $m\cdot\mathbb{R}$. Thus a basic open set in $(\mathbb{R}^n, m\cdot\mathbb{R})$ is

$$\{f \mid f(\Pi [-s_\alpha, s_\alpha]) \subset \Gamma(-t_\beta, t_\beta)\} .$$

If

$$f = \{f_{\alpha\beta}\}$$

then f belongs to that set iff

$$\sum_\beta |\sum_\alpha f_{\alpha\beta} s_\alpha| \Big/ |t_\beta| < 1 .$$

Now let

$$r_{\alpha\beta} = 2^{-\alpha-\beta} t_\beta/s_\alpha .$$

Then supposing

$$\{f_{\alpha\beta}\} \in \Gamma(-r_{\alpha\beta}, r_{\alpha\beta}) \subset (n \times m)\cdot\mathbb{R} ,$$

we have,

$$\sum_\beta | \sum_\alpha f_{\alpha\beta} s_\alpha| / |t_\beta| \le \sum\sum|f_{\alpha\beta} s_\alpha/t_\beta|$$

$$< \sum\sum 2^{-\alpha-\beta} = 1 .$$

Thus the map is continuous and hence a homeomorphism.

The failure of such an argument for larger than countable exponents is the reason for the restriction to finite and countable exponents. Without this, the topology on (C,D) is not a convergence topology which hypothesis figures crucially in several places. Thus I do not know whether the main results of [Barr, 1977] are correct without restriction to at most countable exponents.*

The remaining cases are handled by trivial modification of one of the above arguments and are left to the reader.

(5.32) <u>Proposition</u>. For any $C \in \underline{C}$, $D \in \underline{D}$,

$$(C, D) \cong (D^*, C^*) .$$

Proof. Just examine the nine cases individually. Each of them is trivial in view of the definitions.

---

* The error is in the proof of 7.9 in which the *caveat* of 5.3 is ignored.

(5.33) Proposition. For any $C_1, C_2 \in \underline{C}$, $D \in \underline{D}$,

$$(C_1, (C_2,D)) \cong (C_2, (C_1,D)) \quad .$$

In particular, there is a 1-1 correspondence between maps

$$C_1 \longrightarrow (C_2, D)$$

and maps

$$C_2 \text{---} (C_1, D) \; .$$

Proof. There are, in principal, 27 cases to consider. However, using the symmetry above and the fact that some are zero, there is a considerable collapsing. If C_1 and C_2 are compact and D discrete, both $(C_1,(C_2,D))$ and $(C_2,(C_1,D))$ are discrete. so that we need only show they have the same elements. Now a map

$$f : C_1 \longrightarrow (C_2, D)$$

is a map of a compact to discrete set and hence has a finite image. It thus determines a finite number of continuous maps

$$C_2 \longrightarrow D$$

each of which has open kernel. The intersection of the finitely many open subgroups is an open subgroup which is the kernel of the induced map

$$C_2 \longrightarrow (|C_1|, D) \quad ,$$

whence that map is continuous. Each element of C_2 determines a map $|C_1| \longrightarrow D$ whose kernel contains the kernel of f. Since that is open, each such map is continuous so we have

$$C_2 \longrightarrow (C_1, D) \quad .$$

It is evidently exactly the same in the other direction. Since exponents in the first variable and coefficients in the second come out as coefficients, it is sufficient to consider the cases C_i compact or $\mathbb{R}$ or $\mathbb{Z}$ and D discrete or $\mathbb{R}$ or T. If C_1, is compact, $C_2 = \mathbb{R}$, there are no non-zero maps

$$C_1 \longrightarrow (C_2, D)$$

since (C_2,D) is an $\mathbb{R}$-vector space. On the other hand, (C_1,D) is discrete no matter what choice of D in $\underline{D}$, so also

$$(C_2, (C_1, D)) = 0 \quad .$$

Exactly the same is the case C_2 compact, $C_1 = \mathbb{R}$. The cases $C_1 = \mathbb{Z}$ or $C_2 = \mathbb{Z}$ leave nothing to prove. It is left to the reader to show that using (5.32), every case can be reduced to one already considered.

(5.34) This shows that III.4.2(iii) is satisfied. We now turn to (iv), for which it is sufficient to verify the hypotheses of II.2.9. We turn to the injectivity of T as the cosmallness will appear as a way station and the third hypothesis is evident. Suppose

$$A \longrightarrow B$$

is an embedding. Since B can be embedded in a product of D_ω it suffices to show that every $f : A \longrightarrow T$ can be extended to $B \longrightarrow T$ in the case $B = \Pi D_\omega$. The set

$$f^{-1}\left(-\frac{1}{4}, \frac{1}{4}\right)$$

is open in A and hence contains a set of the form

$$A \cap M$$

where M is open in B. The topology on B is such that there is a finite set of indices $\omega_1, \ldots, \omega_n$ such that

$$M \supset B_0 = \Pi \{B_\omega | \omega \neq \omega_1, \ldots, \omega_n\} .$$

Since B_0 is a subgroup so is $A_0 = A \cap B_0$. Since

$$f(A_0) \subset \left(-\frac{1}{4}, \frac{1}{4}\right)$$

and that set contains no subgroups,

$$f(A_0) = 0 .$$

Let A/A_0 denote the abstract quotient group topologized (here only!) as a subgroup of

$$B/B_0 = D_{\omega_1} \times \ldots \times D_{\omega_n} .$$

The homomorphism

$$\bar{f} : A/A_0 \longrightarrow T$$

induced by f is still continuous. In fact, a homomorphism to T is continuous iff the inverse image of $(-1/4, 1/4)$ is open. But M can be chosen to be the inverse image in B of an open set $\bar{M}$ in B/B_0 since such sets are basic neighborhoods of 0 in B. Then

$$\bar{f}^{-1}(-1/4, 1/4) \supset (A/A_0) \cap \bar{M}$$

and so $\bar{f}$ is continuous in the topology induced on A/A_0. Since T is complete, $\bar{f}$ has a unique extension to the closure of A/A_0. Since $\underline{D}$ is closed under finite products, $B/B_0 \in \underline{D}$. Thus it is sufficient to show that if $D \in \underline{D}$ and $A \subset D$ is a closed subgroup every map $A \longrightarrow T$ has an extension to D. We consider first the case that $D = m \cdot \mathbb{R}$. Let A_0 be the sum of all $\mathbb{R}$-subspaces of A. In fact, since A is closed, A_0 is the divisible subgroup of A. Then A_0 is a subspace of D and there is an $\mathbb{R}$-linear retraction $D \longrightarrow A_0$. Since any $\mathbb{R}$-linear map on a finite dimensional space to an abitrary topological vector space is continuous, this map is continuous on every finite dimensional subspace of $m \cdot \mathbb{R}$. Since $m \cdot \mathbb{R}$ is the direct limit of the finite sums (I am tacitly supposing $m = \aleph_0$; if m is finite, there is nothing to prove), the retraction is continuous on D. The restriction to A splits the sequence

$$0 \longrightarrow A_0 \longrightarrow A \longrightarrow A/A_0 \longrightarrow 0$$

which means that A/A_0 is isomorphic to a subgroup of D. Let A_1 be a subspace of D containing A/A_0 such that

$$D \cong A_0 \oplus A_1$$

algebraically. The map $D \longrightarrow A_0 \oplus A_1$ is continuous by the same argument as above while A_0 and A_1 being subspaces implies the continuity in the opposite direction so the above isomorphism is topological as well. Now $A/A_0 \subset A_1$ so that

$$A \cong A_0 \oplus A/A_0 \quad .$$

For every finite subspace $V \subset A_1$, $V \cap A/A_0$ contains no vector subspace and is hence discrete. Thus every subset of A/A_0 is relatively open in every such V and is therefore relatively open in A/A_0 . Thus A/A_0 is discrete and we have $A \in \underline{D}$. The duality of $\underline{C}$ and $\underline{D}$ implies that

$$D^* \longrightarrow A^*$$

is surjective and the result follows for $D = m\cdot\mathbb{R}$.

Next suppose $D = m\cdot\mathbb{R} + n\cdot T$. Consider the pullback

$$\begin{array}{ccc} B & \longrightarrow & (m+n)\cdot\mathbb{R} \\ \downarrow & & \downarrow \\ A & \longrightarrow & m\,\mathbb{R} \oplus n\cdot T \end{array}$$

where the kernel of each vertical map is $n\cdot\mathbb{Z}$. A map $A \longrightarrow T$ is the same as a map $f : B \longrightarrow T$ such that $f(n\mathbb{Z}) = 0$. Such a map has, by the above, an extension to a map

$$g : (m+n)\cdot\mathbb{R} \longrightarrow T$$

for which, evidently, $g(n\cdot\mathbb{Z}) = 0$, and so induces the desired map $D \longrightarrow T$.

Finally, let

$$D = D_1 \oplus m\cdot\mathbb{R} \oplus n.T$$

with D_1 discrete. Let

$$A_0 = A \cap m\cdot\mathbb{R} \oplus n.T$$

(Note that $m.\mathbb{R} + n.T$ is uniquely determined, being the identity component of D.) Then if

$$f : A \longrightarrow T$$

is a map, the restriction

$$f_0 : A_0 \longrightarrow T$$

has an extension to $m.\mathbb{R} + n.T$ and then to a map

$$g : D \longrightarrow T \quad .$$

The map

$$f - g|A : A \longrightarrow T$$

vanishes on A_0 and hence induces a map

$$\bar{f} : A/A_0 \longrightarrow T \quad .$$

Since A/A_0 is a subgroup of D_1 , that is discrete and T is injective $\bar{f}$ has a continuous extension to a map $\bar{h} : D_1 \longrightarrow T$. Composed with the projection $D \longrightarrow D_1$ we get a map

$$h : D \longrightarrow T$$

such that

$$h|A = f - g|A$$

or

$$f = g|A + h|A = (g+h)|A \quad .$$

Thus $g + h$ is the desired extension.

(5.35) To take care of III.4.2(v), we consider the case of $C \in \underline{C}$ and $A \subset C$ a subgroup such that $C^* \longrightarrow A^*$ is injective. This is done by successively enlarging the class of C. Suppose first that

$$C = F \times \mathbb{R}^n$$

where F is a finitely generated abelian group and n is finite. In that case, $C \in \underline{D}$. If A is a proper closed subgroup let $x \in C - A$ and B be the subgroup generated by x and A. Then B/A is cyclic and has a non-zero map to T. The composite

$$B \longrightarrow B/A \longrightarrow T$$

can, by the previous section be extended to C. Thus $C^* \longrightarrow A^*$ is not injective.

The case

$$C = F \times \mathbb{R}^n \times T^m$$

with m and n finite is easily reduced to the previous one by pulling back along $\mathbb{R}^m \longrightarrow T^m$. Rewrite that as

$$C = \left(F_0 \oplus T^m\right) \times \mathbb{R}^n \times \mathbb{Z}^k$$

where F_0 is finite. Then $F_0 \oplus T^m$ is a compact subgroup of a finite dimensional torus and is, in fact, the most general such. For the dual of a compact subgroup of a torus is the quotient of a finitely generated free group. This is the direct sum of a free group and finite group and its dual is the sum of a torus and a finite group. Now in the most general case,

$$C = C_1 \times \mathbb{R}^n \times \mathbb{Z}^k .$$

It follows from the duality that C_1 can be embedded in a power T^m where m may be arbitrary. If A is not dense in C, it follows from the definition of the product topology that there are finite sets $m_0 \subset m$, $n_0 \subset n$, $k_0 \subset k$ such that the image A_0 of A is not dense in the image C_0 of C in

$$T^{m_0} \times \mathbb{R}^{n_0} \times \mathbb{Z}^{k_0} .$$

The group C_0 is the product of the image of C_1 in T^{m_0} with $\mathbb{R}^{n_0} \times \mathbb{Z}^{k_0}$. That is,

$$C_0 \cong \left(F_0 \oplus T^{m_0}\right) \times \mathbb{R}^{n_0} \times \mathbb{Z}^{k_0}$$

and A_0 is a non-dense subgroup. The result now follows from the previous case.

(5.36) The last hypothesis follows readily from 5.15. All the hypotheses satisfied, it now follows that the theory applies. It is the case that the duality theory described here extends that of Pontrjazin although not sufficient machinery is developed here to show that. The missing facts are these: (See [Hewitt-Ross], 89, especially 9.6 and 9.8 .)

1. Every locally compact group with a norm (i.e. a seminorm taking on the value zero only at zero) is of the form $D \oplus n\cdot\mathbb{R} \oplus m\cdot T$ where D is discrete and n, m are finite;

2. Every locally compact group generated by a compact set is of the form $C \times \mathbb{R}^n \times \mathbb{Z}^m$ where C is compact and m, n are finite.

For if $\underline{D}_0$ and $\underline{C}_0$ are the full subcategories of $\underline{D}$ and $\underline{C}$ described above, it is clear that $\underline{D}_0$ and $\underline{C}_0$ are dual under the duality. Moreover every locally compact group is a subgroup of a product of groups in $\underline{D}_0$. For if L is such a group and p a seminorm, let

$$L_p = \{x \in L \mid p(x) = 0\}$$

$$L/p = L/L_p \quad .$$

Then

$$L \subset \Pi\{L/p \mid p \text{ a seminorm on } L\}$$

and $L/p \in \underline{D}_0$. Since every locally compact is complete, hence ζ-complete, it follows that they all lie in $\underline{G}$. Let L be a locally compact group and X be a compact set. Let M be a compact neighborhood of 0. Then $X + M$ is a compact set with non-empty interior. The subgroup L_0 generated by $X + M$ also has non-empty interior and is hence open. An open set in a locally compact space is locally compact so $L_0 \in \underline{C}_0$. The set $X \subset L_0$. Hence

$$\{f : L \longrightarrow T \mid f(X) \subset (-\tfrac{1}{4}, \tfrac{1}{4})\}$$

is forced to be open by the map

$$L^* \longrightarrow L_0^* \quad .$$

This shows that L^* has the compact/open topology.

(5.37) It is not true, by the way, that every dual has the compact/open topology. To see this, let A be the group of integers topologized as a subgroup of

$$\mathbb{Z}_2 \times \mathbb{Z}_4 \times \mathbb{Z}_8 \times \cdots$$

which is a subgroup of $T^{\aleph_0}$. The maps $A \longrightarrow T$ are found among the maps $\mathbb{Z} \longrightarrow T$, i.e. the elements of T. A^* consists of all the elements of T which annihilate $2^n\mathbb{Z}$ for all sufficiently large n. That is, they are the elements of T whose order is a power of 2. In A the sequence

$$2, 4, 8, 16, \ldots, 2^n, \ldots$$

converges to 0, so that the set

$$X = \{0, 2, 4, 8, 16, \ldots, 2^n, \ldots\}$$

is compact. Suppose

$$f : A \longrightarrow T$$

is a homomorphism such that

$$f(X) \subset (-\tfrac{1}{4}, \tfrac{1}{4}) \quad .$$

Then I claim that $f = 0$. For if not, let ε be the absolutely least residue of $f(2)$ If $\varepsilon > 0$, there is a k such that

$$\tfrac{1}{4} \le 2^k \varepsilon < \tfrac{1}{2}$$

and then $f(2^k) \not\in (-1/4, 1/4)$. Thus $f(2) = 0$ from which $f(2^k) = 0$ for all k . Using the binary representation of n , we see that $f(n) = 0$ as well. This means that the zero homomorphism is open in the compact open topology on A^* . Thus A^* is the group $\mathbb{Z}_{2^\infty}$ of elements of 2 power order, topologized discretely. In fact

$$A^* = \varinjlim \mathbb{Z}_{2^n} .$$

Then

$$A^{**} = \varprojlim \mathbb{Z}_{2^n}{}^* = \varprojlim \mathbb{Z}_{2^n}$$

which is the 2-adic completion of $\mathbb{Z}$.

6. Semilattices

(6.1) By an inf semilattice is meant a partially ordered set in which any finite set of elements has an inf. Obviously it is sufficient that the empty set as well every pair of elements have an inf. If the empty inf is denoted 1 and the inf of x and y by xy we see that an inf semilattice is exactly the same as a commutative monoid in which every element is idempotent. If L_1 and L_2 are semilattices a morphism $f : L_1 \longrightarrow L_2$ is a function which preserves 1 and preserves finite inf, i.e. a monoid homomorphism. We let $\underline{L}$ denote the category of these semi-lattices.

(6.2) If f and g are homomorphisms, so is the map fg defined by

$$fg(x) = f(x)g(x) .$$

The map $u : L_1 \longrightarrow L_2$ such that $u(x) = 1$ for all $x \in L_1$ is a homomorphism and obviously a unit for the multiplication defined above. Thus the set of maps $L_1 \longrightarrow L_2$ is an object $\underline{L}(L_1,L_2)$ of $\underline{L}$. It is evident, since the internal hom is computed pointwise, that the theory of $\underline{L}$ is commutative and hence that $\underline{L}$ is an autonomous variety. Thus parts i) and ii) of III.4.2 are satisfied.

(6.3) We take for $\underline{D}$ the subcategory of Un $\underline{L}$ consisting of all the uniformly discrete semilattices. For $\underline{C}$ we take all the compact semilattices which can be embedded in a product of topologically discrete ones (or, ultimately, of 2 element ones).

(6.4) We let $I = T$ be the two element lattice $\{0,1\}$ with $0 < 1$. If $C \in \underline{C}$, we let C^* be the set of continuous maps $C \to I$ with the discrete uniformity and the lattice structure of $\underline{L}(|C|, I)$. If $D \in \underline{D}$ we let D^* be the lattice, $\underline{L}(D,I)$, equipped with the uniformity induced by I^D. Since I is compact, so is I^D and one easily sees, by the usual arguments, that $\underline{L}(D,I)$ is closed, hence compact. If D is discrete and $x \in D$, the function

$$\hat{x} : D \longrightarrow T$$

defined by

$$\hat{x}(y) = \begin{cases} 1 , \text{ if } x \quad y \\ 0 , \text{ otherwise} \end{cases}$$

is readily seen to preserve inf. From this, it is immediate that every discrete semilattice and hence every object of $\underline{C}$ has enough representations into T and that the canonical maps

$$D \longrightarrow D^{**}$$
$$C \longrightarrow C^{**}$$

are injections.

(6.5) To see they are isomorphisms, we proceed as follows. Let L be a finite lattice. Then given

$$\varphi : L \longrightarrow T$$

let $x = \inf\{y \in L | \varphi(y) = 1\}$. Since L is finite and φ preserves finite inf, $\varphi = \hat{x}$. It is immediate that $x \le y$ implies $\hat{y} \le \hat{x}$ so that L^* is simply the lattice L^{op} . (Of course a <u>finite</u> semilattice is a lattice. However remember that maps preserve only infs. If

$$f : L_0 \longrightarrow L_1$$

is an inf preserving morphism, the induced

$$f^* : L_1^{op} \longrightarrow L_0^{op}$$

also preserves infs. The corresponding function $f^{*op}: L_1 \longrightarrow L_0$ preserves sups of course and is actually the left adjoint of f!) Thus in this case the duality is clear. An arbitrary discrete semilattice is the direct limit of finite ones. If

$$L = \varinjlim L_\alpha ,$$

L_α finite, $L^* = \varprojlim L_\alpha^*$ equipped with the inverse limit topology. If

$$\varphi : L^* \longrightarrow 2$$

is uniform (or even continuous) the inverse image of 1 is open in $L^* \subset \pi L_\alpha^*$. This means it contains a set of the form $L^* \cap M$ where M is the inverse image under projection of a subset of a finite product $L_{\alpha_1} \times \ldots \times L_{\alpha_n}$. The same is true of the inverse image of 0 (with a possibly different finite set of indices). The result is that φ depends only on finite many indices, say $\alpha_1, \ldots, \alpha_m$. Thus there is a factorization of φ

$$L^* \longrightarrow L_0 \xrightarrow{\tilde{\varphi}} 2$$

where $L_0 \subset L_{\alpha_1}^* \times \ldots \times L_{\alpha_m}^*$. The identification of representations on finite lattices implies, in particular, that when $L_0 \rightarrowtail L_1$ is an injection, $L_1^* \longrightarrow L_0^*$ is a surjection. In particular there is a map

$$\psi : L_{\alpha_1}^* \times \ldots \times L_{\alpha_m}^* \longrightarrow T$$

which extends $\tilde{\varphi}$. Now we have shown that

$$\Sigma L_\alpha = \Sigma L_\alpha^{**} \longrightarrow L^{**} ,$$

is a surjection, whence so is $L \longrightarrow L^{**}$.

In the process, we have seen that T is cosmall and that when $L = \varprojlim L_\alpha$, then

$$L^* = \varprojlim L_\alpha^* .$$

But then

$$L^{**} = \varinjlim L_\alpha^{**}$$

and with each L_α reflexive, so is L .

(6.6) This establishes the duality between $\underset{\sim}{C}$ and $\underset{\sim}{D}$. As for the hom, we take (C,D) to consist of the sublattice of $(|C|,D)$ consisting of the uniform morphisms, equipped with the discrete uniformity. The map

$$(C,D) \longrightarrow (D^*,C^*)$$

is the evident one and as for its being uniform, there is nothing to prove. The composite

$$(C,D) \longrightarrow (D^*,C^*) \longrightarrow (C^{**},D^{**}) \cong (C,D)$$

is the identity so the first is injective. Since the second is an instance of the first it is injective as well so that each is an isomorphism. The equivalences

$$(I,D) \cong D$$

$$\mathrm{Hom}(I,(C,D)) \cong \mathrm{Hom}(C,D)$$

are clear. In view of $(C,D) \cong (D^*,C^*)$, I.4.6(iii) follows from

$$(C_1,(C_2,D)) \cong (C_2,(C_1,D)) .$$

Since both sides are (topologically) discrete, it is sufficient to show that

$$\mathrm{Hom}(C_1,(C_2,D)) \cong \mathrm{Hom}(C_2,(C_1,D)) .$$

If

$$f : C_1 \longrightarrow (C_2,D)$$

then the image of f is compact and discrete, hence finite. That is, there is an equivalence relation E on C_1 with only finitely many equivalence classes such that f factors

$$C_1 \longrightarrow C_1/E \longrightarrow (C_2,D) .$$

Let $x_1, \ldots, x_n$ be a set of representatives mod E. Each x_i determines a function

$$f(x_i, -) : C_2 \longrightarrow D$$

which similarly factors

$$C_2 \longrightarrow C_2/E_i \longrightarrow D$$

where E_i is an equivalence relation with only finitely many classes. Since

$$C_2/\cap E_i \longrightarrow \Pi C_2/E_i$$

is injective, there are only finitely many classes mod $(\cap E_i)$. Thus f determines an element of

$$(C_1/E, (C_2/\cap E_i, D))$$

$$\cong (C_2/\cap E_i, (C_1/E, D)) \longrightarrow (C_2, (C_1,D)) ,$$

the exchange possible there because all three are discrete and $\underline{L}$ is an autonomous category. This determines a map

$$(C_1,(C_2,E)) \longrightarrow (C_2,(C_1,E))$$

which is evidently an involution. Thus we have a pre - * - automonous situation.

The discrete uniformity of (C,D) is evidently that of global uniform convergence and is a convergence uniformity. Thus the first three parts of III.4.2 are satisfied.

(6.7) We now consider III.4.2(v). Let $C \in \underline{C}$ and $A \subset C$ be a proper closed sublattice. Since C is profinite there is a finite quotient C_0 of C such that the image A_0 remains proper, else A would be dense. Since $C_0^* \longrightarrow A_0^*$ is surjective and not an isomorphism it cannot be injective whence neither is $C^* \longrightarrow A^*$.

(6.8) We now show that the hypotheses of II.2.9 are satisfied. The first is already shown and the third is evident. For the second, it suffices to show that every diagram of the form

$$\begin{array}{ccc} A & \hookrightarrow & \Pi D_\omega \\ \varphi \downarrow & & \\ T & & \end{array}$$

can be completed. As in 6.5 above there is a finite set of indices $\omega_1, \ldots, \omega_n$ such that φ depends only on those indices. We can thus factor the diagram as

$$\begin{array}{ccc} A & \hookrightarrow & \pi D_\omega \\ \downarrow & & \downarrow \\ A_0 & \hookrightarrow & D_{\omega_1} \times \ldots \times D_{\omega_n} = B_0 \\ \downarrow \varphi & & \\ T & & \end{array}$$

Now A_0 and B_0 are discrete. If $B_0^* \longrightarrow A_0^*$ is not surjective let C be its image.

We have the composite

$$A_0 \cong A_0^{**} \longrightarrow C^* \longrightarrow B_0^{**} \cong B_0$$

is injective, whence the first factor is. But B_0^* is compact so that C is closed and by hypothesis proper in A_0^* and this contradicts the previous paragraph.

The last condition of III.4.2 is immediate and hence we have another model of the theory.

(6.9) The category of complete semilattices and complete homomorphisms provides a model of a $*$-autonomous category that is not constructed in the way described here. (A complete semilattice is in fact a complete lattice. It is understood that the category of complete semilattices is the category whose objects are complete lattices and whose maps are complete inf preserving functions.) It is a closed category, in fact models of commutative theory. The triple $\mathbb{T} = (T, \eta, \mu)$ can be described as follows. For X a set, let

$$TX = 2^X$$

and if $f : X \longrightarrow Y$ is a map, Tf is the right adjoint to $2^f : 2^Y \longrightarrow 2^X$. This means that for $A \subset X$,

$$Tf(A) = \{y | f^{-1}(y) \subset A\}.$$

The unit $\eta : X \longrightarrow TX$ is the singleton map and

$$\mu X : 2^{2^X} \longrightarrow 2^X$$

assigns to a family of subsets their intersection. Thus

$$(X, Y)$$

consists of all inf preserving homomorphisms $X \longrightarrow Y$ made into a complete lattice by the infs in Y. In particular,

$$X^* = (X, 2).$$

Every map $X \longrightarrow 2$ is a limit preserving set-valued functor on X considered as a category. Hence X^* and X have isomorphic underlying sets and the isomorphism is readily seen to be order inverting. Thus $X^* \cong X^{op}$ from which the duality and the $*$-autonomous structure are obvious. It has been incorrectly conjectured that this is a compact category: a $*$-autonomous category such that $I = T$ and such that the natural map

$$X^* \otimes Y \longrightarrow (X, Y)$$

induced by exchanging Y and X in

$$X^* \otimes X \longrightarrow I \longrightarrow (Y, Y)$$

is an isomorphism. Here $X^* \otimes X \longrightarrow I$ is evaluation and $I \longrightarrow (X, X)$ the unit map.

Since $(X,Y) = (X \otimes Y^*)^*$, compactness is equivalent (in a *-autonomous category) to either of the natural maps (whose constructions are easy — given $I = T$)

$$X^* \otimes Y^* \longrightarrow (X \otimes Y)^*$$

$$(X,Y)^* \longrightarrow (X^*,Y^*)$$

being an isomorphism. In the case of lattices, this would imply that (X,Y) and (X^{op},Y^{op}) were dual lattices and, in particular, that (X,Y) and (X^{op}, Y^{op}) have isomorphic underlying sets. I have verified, using an HP67 programmable calculator, that when

$$X = X^{op} =$$

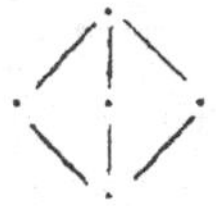

and

$$Y =$$

then there are 94 inf preserving maps $X \longrightarrow Y$ and only 88 such maps $X^{op} \cong X \longrightarrow Y^{op}$. The computation is carried out by modeling Y (as well as Y^{op}) as a set of positive integers ordered by divisibility such that the inf of two numbers is their gcd. This would seem to be possible for any finite lattice.

To have compactness, it is necessary — and almost surely sufficient — to have a trace map

$$tr : (X,X) \cong X^* \otimes X \longrightarrow I ,$$

such that the composite

$$(X,I) \longrightarrow (X,(X,X)) \longrightarrow (X,(X,X)) \xrightarrow{(X,tr)} (X,I)$$

is the identity. Here the first map is induced by the unit and the second interchanges the first and second copies of X. The full subcategory of the category of complete lattices of complete atomic boolean algebras (which is the Kleisli category for the triple) has such a trace map but its form suggests very much that it cannot be extended to any larger subcategory. Namely, if the lattice $X = 2^A$, then let

$$tr : (X,X) \longrightarrow 2$$

be defined by

$$tr(f) = \begin{cases} 1, & \text{if } a \in f(A - \{a\}) \text{ for all } a \in A \\ 0, & \text{otherwise} . \end{cases}$$

In other words, trace is represented by the map $g : X \longrightarrow X$ defined by

$$g(A) = A , \quad g(A-\{a\}) = a , \quad g(A_0) = \emptyset$$

for any subset $A_0 < A$ that omits at least two elements of A. The fact that X is complemented and atomic seems crucial here.

BIBLIOGRAPHY

M. Barr, Duality of vector spaces, Cahiers Topologie Géométrie Différentielle 17 (1976),3-14.

M. Barr, Duality of banach spaces, Cahiers Topologie Géométrie Différentielle 17 (1976), 15-32

M. Barr, Closed categories and topological vector spaces, Cahiers Topologie Géométrie Différentielle 17 (1976), 223-234.

M. Barr, Closed categories and banach spaces, Cahiers Topologie Géométrie Différentielle 17 (1976), 335-342.

M. Barr, A closed category of reflexive topological abelian groups, Cahiers Topologie Géométrie Différentielle 18 (1977), 221-248.

M. Barr, The point of the empty set, Cahiers Topologie Géométrie Différentielle 13 (1973), 357-368.

S. Eilenberg, G.M. Kelly, Closed categories, Proc. Conf. Categorical Algebra (La Jolla, 1965), Springer-Verlag, 1966, 421-562.

E. Hewitt, K.A. Ross, Abstract Hamonic Analysis, Vol. I, 1963, Springer-Verlag.

K.H. Hofmann, M. Mislove, A. Stralka, The Pontryagin Duality of Compact O-Dimensional Semilattices and its Applications, Lecture Notes Math. 396, (1974), Springer-Verlag.

J.R. Isbell, Uniform Spaces, Amer. Math. Soc. Surveys no. 12, 1964.

J.L. Kelley,General Topology, Van Nostrand, 1955.

G.M. Kelly, Monomorphisms, epimorphisms and pull-backs, J. Austral. Math. Soc. 9, (1969), 124-142.

F.W. Lawvere, Functional Semantics of Algebraic Theories, Dissertation, Columbia University, 1963.

S. Lefschetz, Algebraic Topology, Amer. Math. Soc. Colloquium Publications, Vol. XXVII, 1942.

F.E.J. Linton, Some aspects of equational categories, Proc. Conf. Categorical Algebra (La Jolla, 1965), Springer-Verlag, 1966, 84-94.

A. Pietsch, Nuclear Locally Convex Spaces, Springer-Verlag, 1972.

H.H. Schaefer, Topological Vector Spaces, third printing, Springer-Verlag, 1970.

Z.Samadeni, Projectivity, injectivity and duality, Rozprawy Mat. 35 (1963).

A. Wiweger, Linear spaces with mixed topology, Studia Math. 20 (1961), 47-68.

CONSTRUCTING *-AUTONOMOUS CATEGORIES

Po-Hsiang Chu

CHAPTER I: PRELIMINARIES

We will be dealing with closed symmetric monoidal (autonomous) and *-autonomous categories as defined in the previous paper. Using the MacLane-Kelly coherence conditions (see [MacLane,Kelly]), M.F. Szabo has proved the following useful theorem [to appear].

<u>Theorem</u>: A diagram commutes in all closed symmetric monoidal categories iff it commutes in the category of real vector spaces.

This theorem not only points out the notion of closed symmetric monoidal category is a 'correct' generalization of the category of vector spaces, but it also provides a very easy method to check if a diagram is commutative in any closed symmetric monoidal category.

The following is a collection of easy consequences of this theorem which we shall use later on:

<u>Corollary 1.</u> Given A,B,C objects in $\underline{V}$ and map $A\otimes B \xrightarrow{f} C$, then the following diagram commutes:

$$\begin{array}{ccc} I & \xrightarrow{i} & (A,A) \\ {\scriptstyle i}\downarrow & & \downarrow \\ (B,B) & \longrightarrow & (A\otimes B,C) \end{array}$$

where the map $(A,A) \longrightarrow (A\otimes B,C)$ is the composition

$(A,A) \xrightarrow{(id,\tilde{f})} (A,(B,C)) \xrightarrow{p^{-1}} (A\otimes B,C)$

Note. $A \xrightarrow{\tilde{f}} (B,C)$ is the usual transpose of $A\otimes B \xrightarrow{f} C$.

The map $(B,B) \longrightarrow (A\otimes B,C)$ is obtained in a similar fashion. From now on we simply denote either composite by $\tilde{f}$.

<u>Corollary 2</u>. Given A,B,C,D,F objects in $\underline{V}$ and map $B\otimes C \longrightarrow F$, then the following diagram commutes:

$$\begin{array}{ccc} (A,C)\otimes(D,B) & \xrightarrow{(id,f)\otimes if} & (A,(B,F))\otimes(D,B) \\ \downarrow{\scriptstyle id\otimes(id,f)} & & \downarrow{\scriptstyle p^{-1}\otimes id} \\ (A,C)\otimes(D,C,F)) & & (A\otimes B,F)\otimes(D,B) \\ \downarrow{\scriptstyle id\otimes p^{-1}} & & \downarrow{\scriptstyle (s,id)\otimes id} \\ (A,C)\otimes(D\otimes C,F) & & (B\otimes A,F)\otimes(D,B) \\ \downarrow{\scriptstyle id\otimes(s,id)} & & \downarrow{\scriptstyle p\otimes id} \\ (A,C)\otimes(C\otimes D,F) & & (B,(A,F))\otimes(D,B) \\ \downarrow{\scriptstyle id\otimes p} & & \downarrow{\scriptstyle M} \\ (A,C)\otimes(C,(D,F)) & & (D,(A,F)) \\ \downarrow{\scriptstyle s} & & \\ (C,(D,F))\otimes(A,C) & & \downarrow{\scriptstyle p^{-1}} \\ \downarrow & & \\ (A,(D,F)) \xrightarrow{p^{-1}} (A\otimes D,F) & \xrightarrow{(s,id)} & (D\otimes A,F) \end{array}$$

PROOF. It is easy to check that the diagram commutes in the category of real vector spaces.

Remark. The word "coherence" is going to appear frequently throughout this paper. In particular, if the commutativity of a certain diagram is said to be implied by coherence, we understand that its commutativity follows easily from this theorem.

Our second assumption on $\underline{V}$ is that it has pullbacks. Since almost all interesting examples of closed symmetric monoidal categories have this property, this restriction is not too drastic.

The following is a collection of examples satisfies our assumption:

(i) The category of vector spaces over a fixed field K;

(ii) The category of Banach spaces;

(iii) The category of compactly generated spaces;

(iv) The category of sets (and functions);

(v) The category of abelian groups;

(vi) The category of lattices.

An example of a closed symmetric monoidal category that does not have pullbacks is the category of sets and relations.

CHAPTER II: CONSTRUCTION OF $\underline{A}_X$ AND ITS ENRICHMENT OVER $\underline{V}$.

1. The Category $\underline{A}_X$

Given an arbitrary object X in $\underline{V}$, we shall construct a category $\underline{A}_X$ as follows:

The objects of $\underline{A}_X$ consist of triplets (V,V',v) where V,V' are objects in $\underline{V}$ and $v:V\otimes V' \longrightarrow X$ is a morphism in $\underline{V}$.

A morphism from (V,V',v) to (W,W',w) is a pair (f,g), where $f:V \longrightarrow W$ and $g:W' \longrightarrow V'$ are morphisms in $\underline{V}$ such that the square

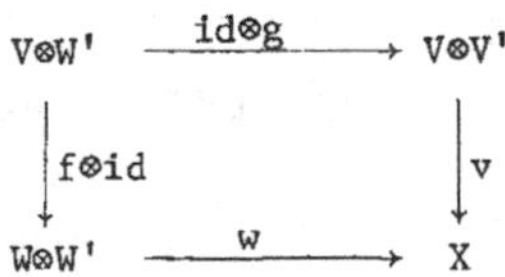

commutes.

If $(f,g):(V,V',v) \longrightarrow (W,W',w)$ and $(h,k):(W,W',w) \longrightarrow (U,U',u)$ are morphisms in $\underline{A}_X$ then the following diagram commutes:

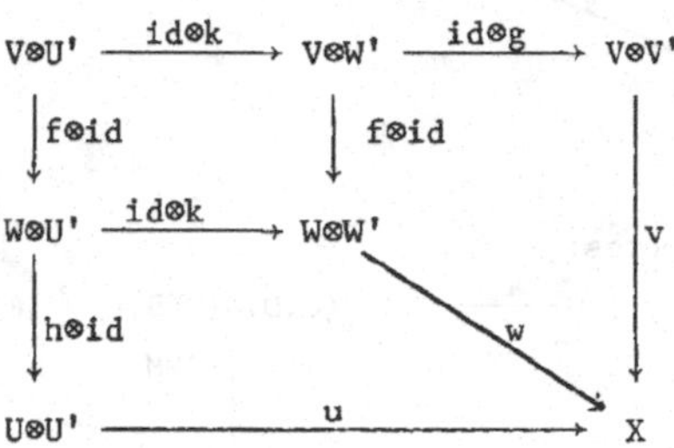

This implies the composition of (f,g) and (h,k) is $(h\circ f, g\circ k)$ in $\underline{A}_X$. Since the composition is defined explicitly in terms of morphisms in $\underline{V}$, the associativity of maps in $\underline{A}_X$ can now be verified:

If $(f,g):(V,V',v) \longrightarrow (W,W',w)$
$(h,k):(W,W',w) \longrightarrow (U,U',u)$
$(1,m):(U,U',u) \longrightarrow (T,T',t)$

are morphisms in $\underline{A}_X$, then

$$\begin{aligned}((1,m)\circ(h,k))\circ(f,g) &= (1\circ h, k\circ m)\circ(f,g)\\ &= ((1\circ h)\circ f,\ g\circ(k\circ m))\\ &= (1\circ(h\circ f),(g\circ k)\circ m)\\ &= (1,m)\circ(h\circ f, g\circ k)\\ &= (1,m)\circ((h,k)\circ(f,g)) .\end{aligned}$$

Moreover, $Id(V,V',v) = (id_V, id_{V'})$ is the obvious identity. Hence we have shown that $\underline{A}_X$ is a category.

2. $\underline{A}_X$ is Enriched over $\underline{V}$

Definition. If $\underline{V}$ is a closed monoidal category, then $\underline{A}$ is enriched over $\underline{V}$ if $\underline{A}$ is equipped with the following:

i) For each A,B in $\underline{A}$, an object $\underline{V}(A,B)$ in $\underline{V}$;
ii) For each A in $\underline{A}$, a morphism $j(A):I \longrightarrow \underline{V}(A,A)$ in $\underline{V}$;
iii) For each A,B,C in $\underline{A}$, a morphism
$M'(A,B,C):\underline{V}(B,C)\otimes\underline{V}(A,B) \longrightarrow \underline{V}(A,C)$ in $\underline{V}$.

These data are required to satisfy the following axioms:

VC 1. The following diagram commutes:

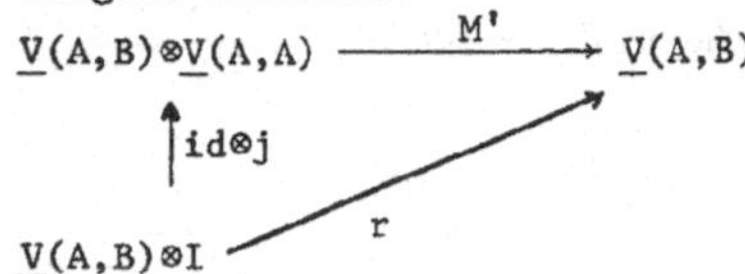

VC 2. The following diagram commutes:

$$\begin{array}{ccc} \underline{V}(A,A)\otimes\underline{V}(B,A) & \xrightarrow{M'} & \underline{V}(B,A) \\ \uparrow{\scriptstyle j\otimes id} & \nearrow{\scriptstyle l} & \\ I\otimes\underline{V}(B,A) & & \end{array}$$

VC 3. The following diagram commutes:

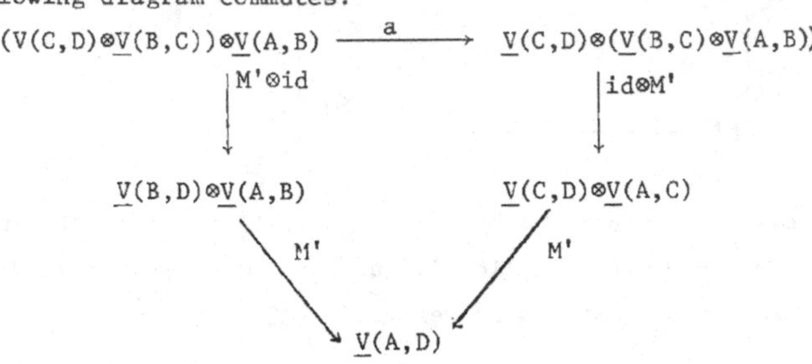

Given $A = (V,V',v), B = (W,W',w)$ objects in $\underline{A}_X$, define $\underline{V}(A,B)$ to be the object in $\underline{V}$ such that the following square is a pullback.

$$\begin{array}{ccc} \underline{V}(A,B) & \xrightarrow{p1} & (V,W) \\ \downarrow{\scriptstyle p2} & & \downarrow \\ (W',V') & \longrightarrow & (V\otimes W',X) \end{array}$$

Here

$$(V,W) \longrightarrow (V\otimes W',X) \quad = \quad (V,W) \xrightarrow{w} (V,(W',X)) \xrightarrow{p^{-1}} (V\otimes W',X)$$

and

$$(W',V') \longrightarrow (V\otimes W',X) \quad = \quad (W',V') \xrightarrow{v} (W',(V,X)) \xrightarrow{p^{-1}} (W'\otimes V,X) \xrightarrow{s} (V\otimes W',X)$$

are the right and bottom maps, respectively. Therefore $\underline{V}(A,B)$ is defined up to isomorphism in $\underline{V}$.

Given $A = (V,V',v)$ in $\underline{A}_X$, the following diagram commutes, by Corollary 1:

$$\begin{array}{ccc} I & \xrightarrow{i} & (V,V) \\ \downarrow{\scriptstyle i} & \text{p.b.} & \downarrow{\scriptstyle \tilde{v}} \\ (V',V') & \xrightarrow{\tilde{v}} & (V\otimes V',X) \end{array}$$

Universal property of pullbacks implies that there exists a unique map $j(A)$ $I \longrightarrow \underline{V}(A,A)$ in $\underline{V}$ such that the diagram

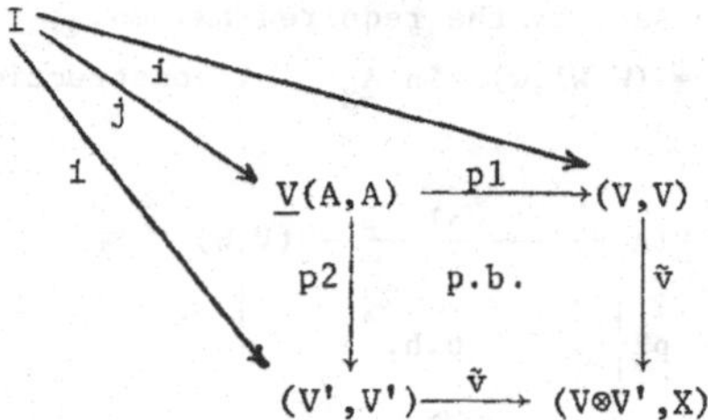

commutes.

Now suppose $A = (V,V',v)$, $B = (W,W',w)$, $C = (T,T',t)$ are three objects in $\underline{A}_X$. In order to verify iii) it suffices to show the outer square of the diagram.

$$\begin{array}{ccccc}
\underline{V}(B,C)\otimes\underline{V}(A,B) & & \xrightarrow{\quad p1\otimes p1 \quad} & & (W,T)\otimes(V,W) \\
\downarrow{\scriptstyle p2\otimes p2} & & & & \downarrow{\scriptstyle M} \\
(T',W')\otimes(W',V') & & \underline{V}(A,C) & \xrightarrow{\ p1\ } & (V,T) \\
\downarrow{\scriptstyle s} & & \downarrow{\scriptstyle p2} & \text{p.b.} & \downarrow{\scriptstyle \tilde{t}} \\
(W',V')\otimes(T',W') & \xrightarrow{\ M\ } & (T',V') & \xrightarrow{\ \tilde{v}\ } & (V\otimes T',X)
\end{array}$$

commutes.

Using the fact that $-\otimes-$ is a bifunctor and

$$\begin{array}{ccc}
\underline{V}(A,B) & \xrightarrow{\ p1\ } & (V,W) \\
\downarrow{\scriptstyle p2} & & \downarrow{\scriptstyle \tilde{w}} \\
(W',V') & \xrightarrow{\ \tilde{w}\ } & (V\otimes W',X)
\end{array}
\qquad\qquad
\begin{array}{ccc}
\underline{V}(B,C) & \xrightarrow{\ p1\ } & (W,T) \\
\downarrow{\scriptstyle p2} & & \downarrow{\scriptstyle \tilde{t}} \\
(T',W') & \xrightarrow{\ \tilde{w}\ } & (W\otimes T',X)
\end{array}$$

are pullbacks (hence commute!), we can get the desired result from the commutative diagram in Fig. 1. Note in Fig. 1 that corollary 2 of Szabo's theorem (Chapter I) implies that (2) commutes; coherence implies that (1) and (3) commute. Again using the universal property of pullbacks, there exists a unique morphism $M'(A,B,C):\underline{V}(B,C)\otimes\underline{V}(A,B) \longrightarrow \underline{V}(A,C)$ in $\underline{V}$ such that the diagram

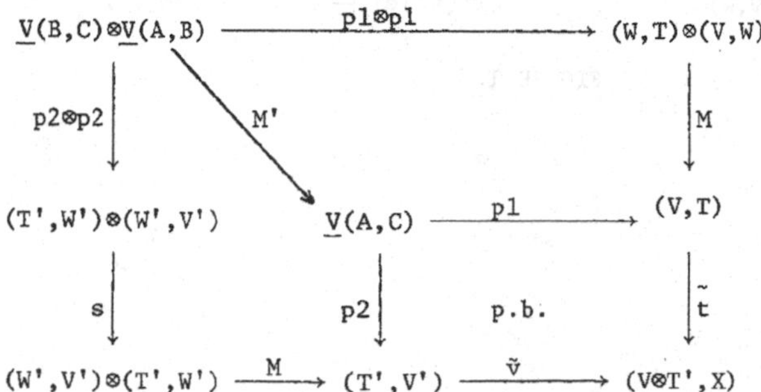

commutes. Hence i) - iii) are defined.

Now we have to show they satisfy the required axioms.

Given $A = (V,V',v)$, $B = (W,W',w)$ in $\underline{A}_X$, by construction we have the pullback diagram:

$$\begin{array}{ccc} \underline{V}(A,B) & \xrightarrow{p1} & (V,W) \\ {\scriptstyle p2}\downarrow & \text{p.b.} & \downarrow{\scriptstyle \tilde{w}} \\ (W',V') & \xrightarrow{\tilde{v}} & (V\otimes W',X) \end{array}$$

But the coherence of $\underline{V}$ implies that the diagrams of Fig. 1 commute.

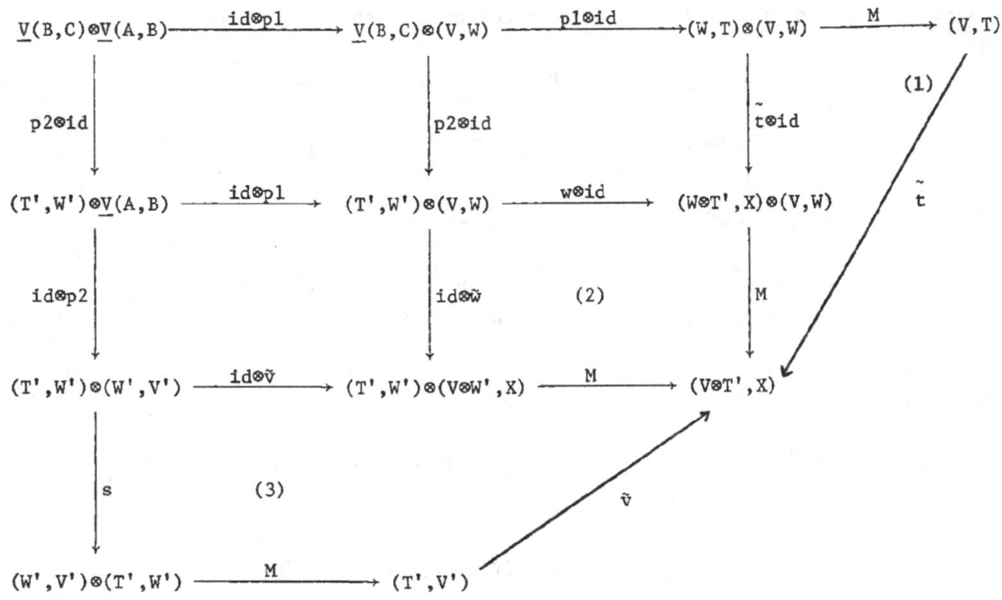

$$\begin{array}{ccc} \underline{V}(A,B)\otimes I & \xrightarrow{r} & \underline{V}(A,B) \\ {\scriptstyle p1\otimes id}\downarrow & & \downarrow{\scriptstyle p1} \\ (V,W)\otimes I & \xrightarrow{r} & (V,W) \end{array} \qquad \begin{array}{ccc} \underline{V}(A,B)\otimes I & \xrightarrow{r} & \underline{V}(A,B) \\ {\scriptstyle p2\otimes id}\downarrow & & \downarrow{\scriptstyle p2} \\ (W',V')\otimes I & \xrightarrow{r} & (W',V') \end{array}$$

FIGURE 1.

Hence the following diagram

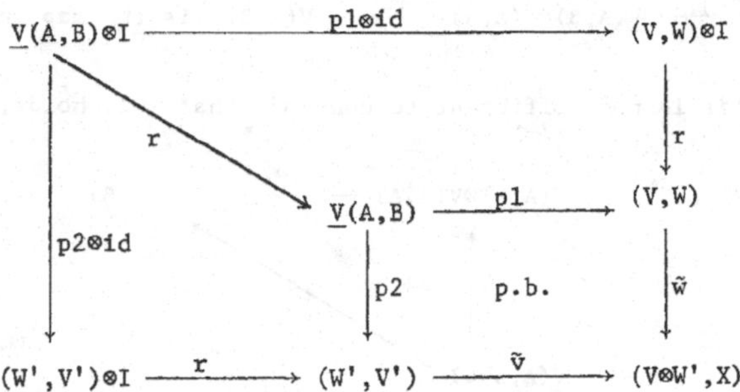

commutes.

Since the outer square commutes, there exists a unique map $\underline{V}(A,B)\otimes I \xrightarrow{r'} \underline{V}(A,B)$ such that (1) and (2) commute. But the map $\underline{V}(A,B)\otimes I \xrightarrow{r} \underline{V}(A,B)$ has this property as well; therefore it follows from uniqueness that it is the map induced by pulling back.

Recall that in the construction of j(A) we have the following commutative diagram:

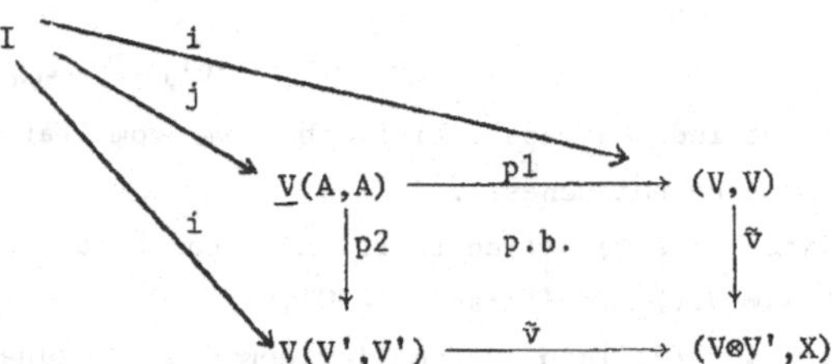

Then the defining property of M(A,B,A), coherence of $\underline{V}$, and the fact that $-\otimes-$ is a bifunctor imply that the diagram:

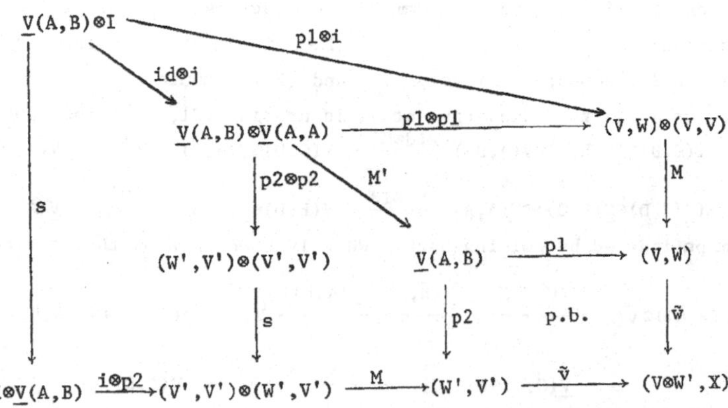

commutes.

Again applying the same argument, we conclude that the map $\underline{V}(A,B)\otimes I \xrightarrow{id\otimes j} \underline{V}(A,B)\otimes\underline{V}(A,A) \xrightarrow{M'} \underline{V}(A,B)$ is the map induced by pulling back.

But this is not sufficient to conclude that VC1. holds, i.e. that the diagram:

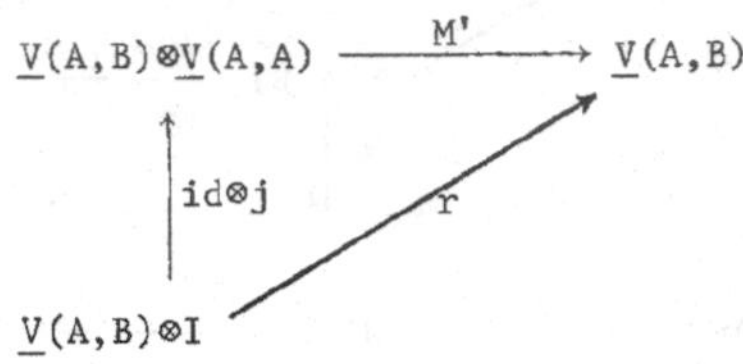

commutes.

We are still required to show that the following diagrams commute:

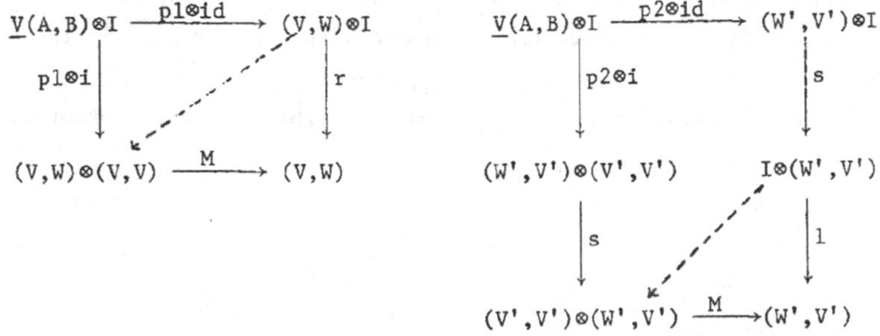

That is, that the induced maps satisfy the same commutative square (therefore they are same by uniqueness).

But it is trivial once we notice there exist canonical maps $(V,W)\otimes I \xrightarrow{id\otimes i} (V,W)\otimes(V,V)$ in (1) and $I\otimes(W',V') \xrightarrow{i\otimes id} (V',V')\otimes(W',V')$ in (2) which break (1) and (2) into two smaller commutative squares. Hence VC1. holds.

Applying a similar argument, we conclude VC2. is also true. Next we are going to verify VC3.

Given $A = (V,V',v)$, $B = (W,W',w)$, $C = (T,T',t)$, $D = (U,U',u)$ objects in $\underline{A}_X$, then by iii) we have the commutative diagrams of Figure 2.

Coherence of $\underline{V}$ and property of $M(A,B,C)$ imply that subdiagrams (1) and (2) of Figure 2 commute; similarly (1') and (2') commute.

Now we apply the same argument as in proving VC1, i.e. the maps

$$\underline{V}(C,D)\otimes(\underline{V}(B,C)\otimes\underline{V}(A,B)) \xrightarrow{id\otimes M'} \underline{V}(C,D)\otimes\underline{V}(A,C) \xrightarrow{M'} \underline{V}(A,D)$$

$$(\underline{V}(C,D)\otimes\underline{V}(B,C))\otimes\underline{V}(A,B) \xrightarrow{M'\otimes id} \underline{V}(B,D)\otimes\underline{V}(A,B) \xrightarrow{M'} \underline{V}(A,D)$$

are the maps induced by pulling back. We only have to show that the composition

$$(\underline{V}(C,D)\otimes\underline{V}(B,C))\otimes\underline{V}(A,B) \xrightarrow{a(\underline{V}(C,D),\ \underline{V}(B,C),\ \underline{V}(A,B))} \underline{V}(C,D)\otimes(\underline{V}(B,C)\otimes\underline{V}(A,B))$$

$$\xrightarrow{id\otimes M'} \underline{V}(C,D)\otimes\underline{V}(A,C) \xrightarrow{M'} \underline{V}(A,D)$$

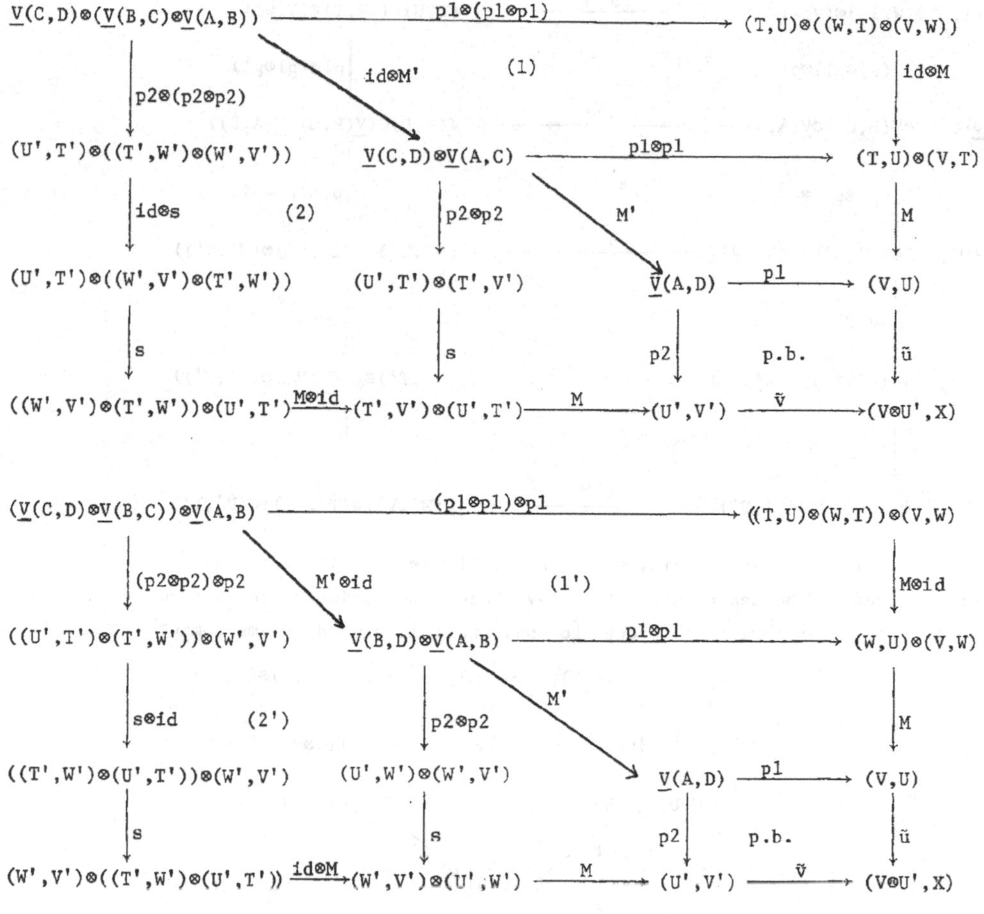

FIGURE 2.

is also a map induced by pullback and it satisfies the same commutative square as the map:

$$(\underline{V}(C,D)\otimes\underline{V}(B,C))\otimes\underline{V}(A,B) \xrightarrow{M'\otimes id} \underline{V}(B,D)\otimes\underline{V}(A,B) \xrightarrow{M'} \underline{V}(A,D)$$

The first part follows easily from the following commutative diagram:

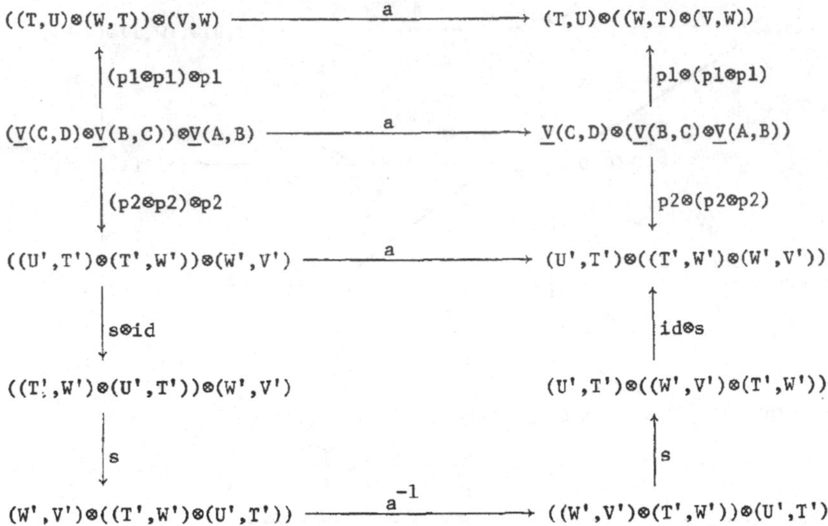

As for the second part, we observe a simple fact of $\underline{V}$: two permutations of the tensor product of any three fixed objects are coherently isomorphic. Therefore it is enough to show the following diagrams commute:

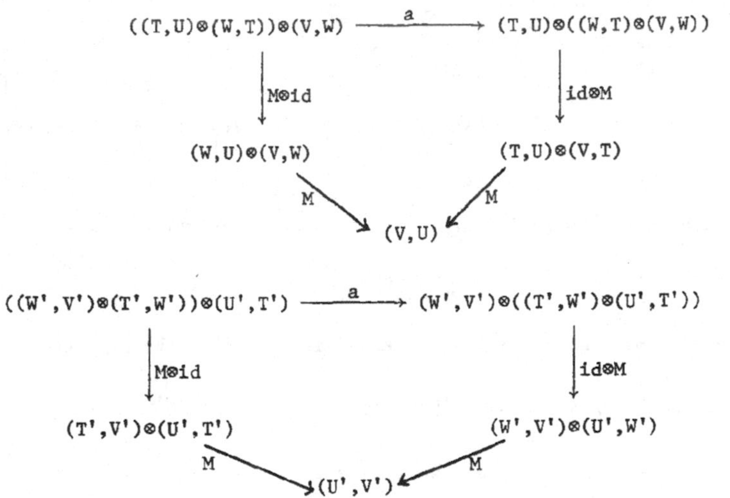

This follows trivially from coherence and completes the proof.

CHAPTER III: $\underline{A}_X$ HAS A *-AUTONOMOUS STRUCTURE

1. The Hom-Functor $\underline{A}_X(-,-)$

Definition. Given any two objects $A = (V,V',v)$ and $B = (W,W',w)$ in $\underline{A}_X$, define an object $\underline{A}_X(A,B) = (\underline{V}(A,B),\ V\otimes W',\ n)$ in $\underline{A}_X$ as follows:

First of all recall $\underline{V}(A,B)$ is the object in $\underline{V}$ such that the following diagram is a pullback.

$$\begin{array}{ccc} \underline{V}(A,B) & \xrightarrow{p1} & (V,W) \\ {\scriptstyle p2}\downarrow & \text{p.b.} & \downarrow{\scriptstyle \tilde{w}} \\ (W',V') & \xrightarrow{\tilde{v}} & (V\otimes W',X) \end{array}$$

Since we require $\underline{A}_X(A,B)$ to be an object in $\underline{A}_X$, n has to be a morphism in $\underline{V}$, which sends $\underline{V}(A,B)\otimes(V\otimes W')$ to X.

It seems there are two (canonical) alternatives for defining n:

(1) Since the above square commutes, let n' be the morphism (along either route) which sends $\underline{V}(A,B)$ to $(V\otimes W',X)$, and define $n:\underline{V}(A,B)\otimes(V\otimes W')\longrightarrow X$ to be the transpose of n'.

(2) Again since the above square commutes, we have the following commutative diagram:

$$\begin{array}{ccc} \underline{V}(A,B)\otimes(V\otimes W') & \xrightarrow{p1\otimes id} & (V,W)\otimes(V\otimes W') \\ {\scriptstyle p2\otimes id}\downarrow & & \downarrow{\scriptstyle \tilde{w}\otimes id} \\ (W',V')\otimes(V\otimes W') & \xrightarrow{\tilde{v}\otimes id} & (V\otimes W',X)\otimes(V\otimes W') \end{array}$$

Now let $ev:(V\otimes W',X)\otimes(V\otimes W') \longrightarrow X$ be the evaluation map, then put n'' = ev composed with the above map $\underline{V}(A,B)\otimes(V\otimes W') \longrightarrow (V\otimes W',X)\otimes(V\otimes W')$.

But since $\underline{V}$ is coherent, it is easy to verify that n is identical to n', so these two definitions are same.

For the rest of this section we shall prove $\underline{A}_X(-,-)$ is a bifunctor which sends $\underline{A}_X^{op} \times \underline{A}_X$ to $\underline{A}_X$.

We have to show

i) given any object $B = (W,W',w)$ in $\underline{A}_X$, $F = \underline{A}_X(-,B)$ is a contravariant functor;

ii) $G = \underline{A}_X(B,-)$ is a covariant functor;

iii) Given $A \longrightarrow B$, $C \longrightarrow D$ in $\underline{A}_X$, then the diagram

$$\begin{array}{ccc} \underline{A}_X(B,C) & \longrightarrow & \underline{A}_X(A,C) \\ \downarrow & & \downarrow \\ \underline{A}_X(B,D) & \longrightarrow & \underline{A}_X(A,D) \end{array}$$

commutes.

Recall if $C = (V,V',v)$ and $A = (P,P',p)$ in $\underline{A}_X$ and $(f,g):C \longrightarrow A$ is a morphism in $\underline{A}_X$, then the square:

$$\begin{array}{ccc} V\otimes P' & \xrightarrow{\ id\otimes g\ } & V\otimes V' \\ \Big\downarrow{\scriptstyle f\otimes id} & & \Big\downarrow{\scriptstyle \tilde{v}} \\ P\otimes P' & \xrightarrow{\ \tilde{p}\ } & X \end{array}$$

commutes.

In order to show F is contravariant, we must find a map (in $\underline{A}_X$)

$$F(f,g) = (f',g'):\underline{A}_X(A,B) \longrightarrow \underline{A}_X(C,B) .$$

By definition $\underline{A}_X(A,B) = (\underline{V}(A,B),\ P\otimes W',\ n_1)$ and $\underline{A}_X(C,B) = (\underline{V}(C,B),\ V\otimes W',\ n_2)$; so the choice for g' is clear: $g' = f\otimes id: V\otimes W' \longrightarrow P\otimes W'$

As for f', consider the following diagram:

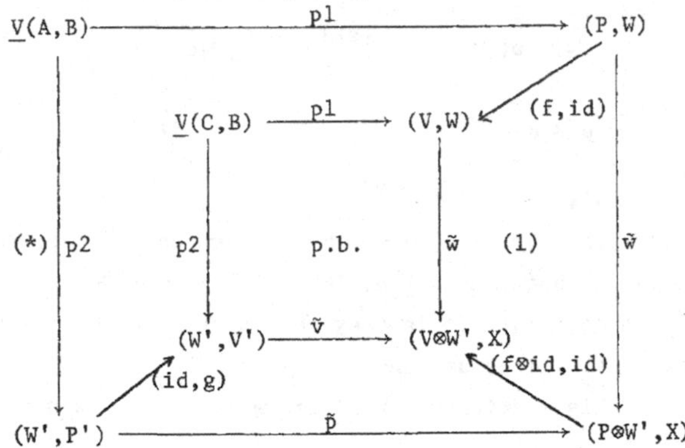

We know the outer square commutes, therefore it suffices to show (1) and (2) are commutative.

For (1), we prove it by looking at the following commutative diagram:

$$\begin{array}{ccccccc} & & (P,W) & \xrightarrow{(f,id)} & (V,W) & & \\ & \swarrow{\scriptstyle \tilde{w}} & & & & \searrow{\scriptstyle \tilde{w}} & \\ (P,(W',X)) & \xrightarrow{p^{-1}} & (P\otimes W',X) & \xrightarrow{(f\otimes id,\ id)} & (V\otimes W',X) & \xleftarrow{p^{-1}} & (V,(W',X)) \end{array}$$

As for (2) we have a similar diagramatical proof:

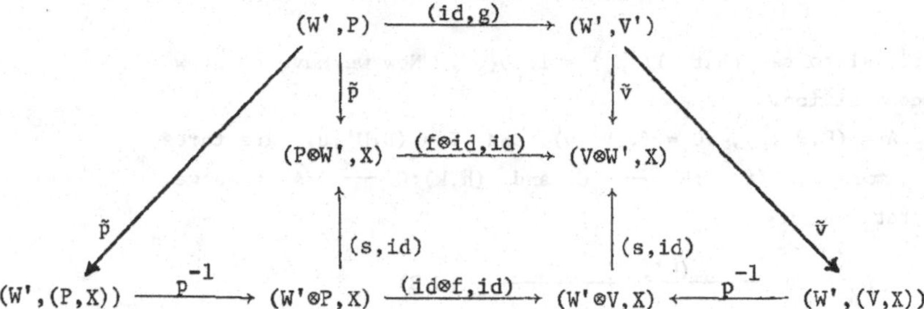

But in this case the commutativity of the outer square is due to the fact that (f,g) is a morphism which sends C to A (hence the diagram (*) above commutes).

This implies that there is a unique map $f':\underline{V}(A,B) \longrightarrow \underline{V}(C,B)$ induced by pullback such that the diagram

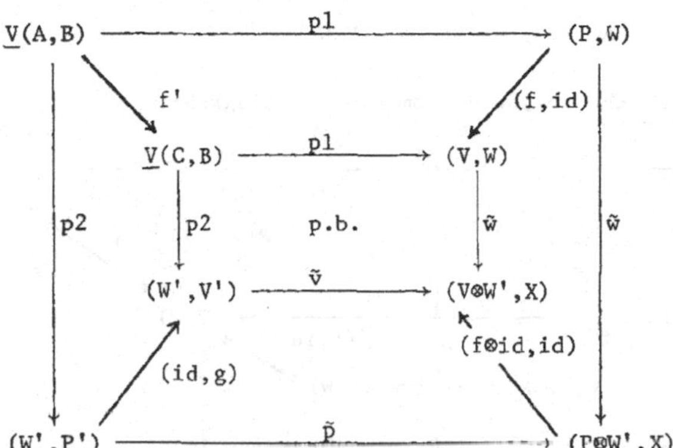

commutes.

Therefore the following diagram commutes:

$$\begin{array}{ccccc} \underline{V}(A,B) & \xrightarrow{p2} & (W',P) & \xrightarrow{p} & (P\otimes W',X) \\ \downarrow{\scriptstyle f'} & & \downarrow{\scriptstyle (id,g)} & & \downarrow{\scriptstyle (f\otimes id,id)} \\ \underline{V}(C,B) & \xrightarrow{p2} & (W',V') & \xrightarrow{v} & (V\otimes W',X) \end{array}$$

This implies that:

$$\begin{array}{ccc} \underline{V}(A,B)\otimes(V\otimes W') & \xrightarrow{id\otimes g'} & \underline{V}(A,B)\otimes(P\otimes W') \\ \downarrow{\scriptstyle f'\otimes id} & & \downarrow{\scriptstyle n_1} \\ \underline{V}(C,B)\otimes(V\otimes W') & \xrightarrow{n_2} & X \end{array}$$

commutes.

Therefore (f',g') has the property required of a morphism in $\underline{A}_X$.

It is trivial to see that $F(id_A) = id_{F(A)}$. Now we have to show F preserves composition.

Suppose $A = (P,P',p)$, $C = (V,V',v)$ and $E = (U,U',u)$ are three objects in $\underline{A}_X$, moreover $(f,g):E \longrightarrow C$ and $(h,k):C \longrightarrow A$ then we want to show that

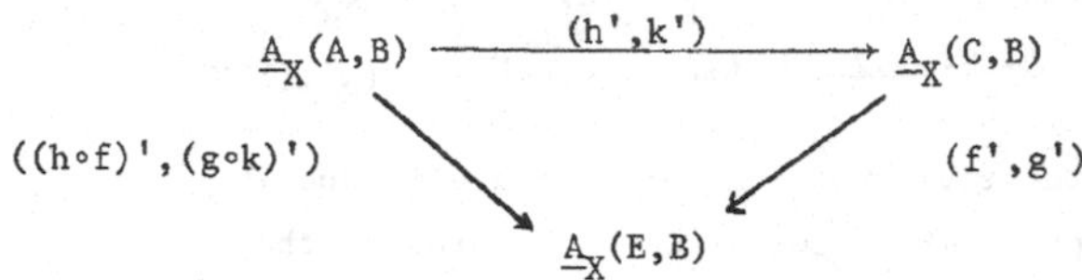

commutes.

By definition: $\underline{A}_X(A,B) = (\underline{V}(A,B),\ P\otimes W',\ n_1)$

$\underline{A}_X(C,B) = (\underline{V}(C,B),\ V\otimes W',\ n_2)$

$\underline{A}_X(E,B) = (\underline{V}(E,B),\ U\otimes W',\ n_3)$

Now we consider the following commutative diagram:

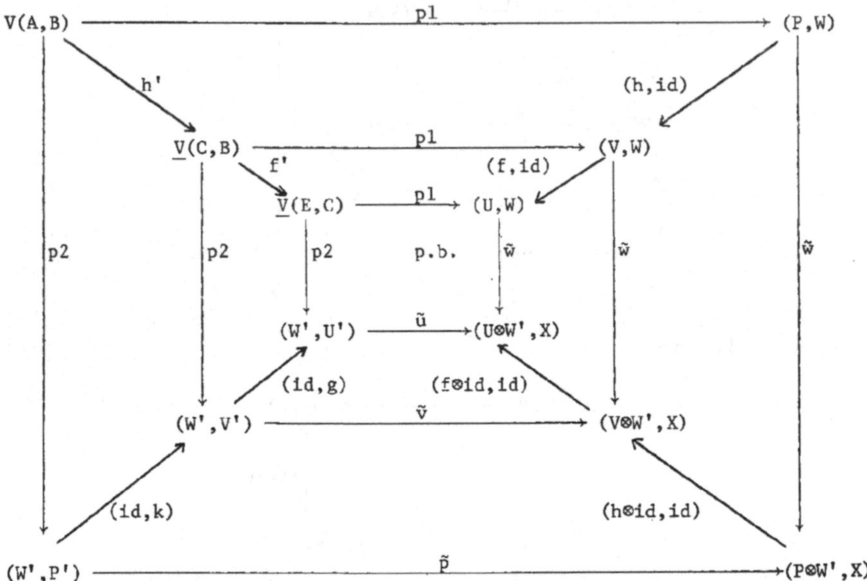

But the following diagrams also commute:

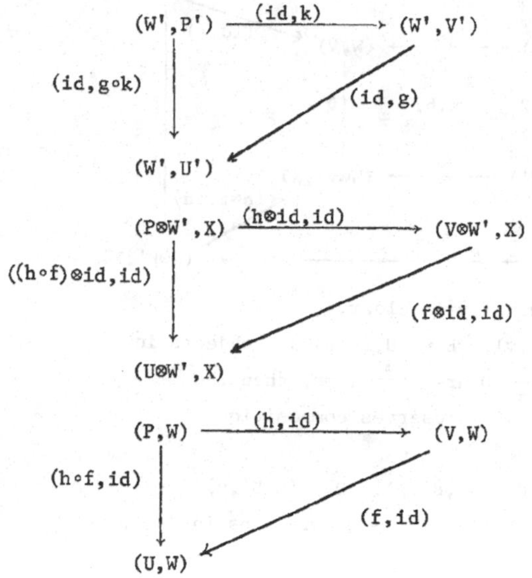

This implies that the map induced by pullback is identical to $f' \circ h'$, and clearly $k' \circ g' = (h \otimes id) \circ (f \otimes id) = ((h \circ f) \otimes id) = (g \circ k)'$. Hence F is a contravariant functor.

As for G, we have a similar series of diagrammatical proofs: Suppose $B = (W,W',w)$, $A = (P,P',p)$, $C = (V,V',v)$ are objects in $\underline{A}_X$ with $(f,g)\colon A \longrightarrow C$ a morphism in $\underline{A}_X$. We need $G(f,g) = (f',g')\colon G(A) \longrightarrow G(C)$. By definition

$$G(A) = \underline{A}_X(B,A) = (\underline{V}(B,A)\ ,\ W \otimes P',\ n_1)$$

and

$$G(C) = \underline{A}_X(B,C) = (\underline{V}(B,C)\ ,\ W \otimes V',\ n_2)\quad .$$

Hence the choice of

$$g' = id \otimes g\colon W \otimes V' \longrightarrow W \otimes P'$$

is clear. And the following commutative diagram shows the existence and uniqueness of f':

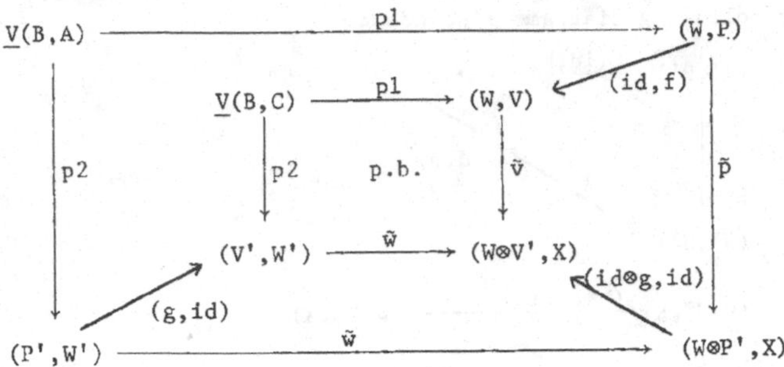

Again the preservation of the identity is clear.

Now if $A = (P,P',p)$, $C = (V,V',v)$, $E = (U,U',u)$ are objects in $\underline{A}_X$ and $(f,g):A \longrightarrow C$, $(h,k):C \longrightarrow D$ are morphisms, then the commutative diagrams of Figure 3 imply G preserves composition.

To prove (iii):

Suppose $A = (V,V'v)$, $B = (W,W',w)$, $C = (P,P',p)$, $D = (U,U',u)$ are objects in $\underline{A}_X$ and $(f,g):A \longrightarrow B$, $(h,k):C \longrightarrow D$ are maps in $\underline{A}_X$, then the following diagrams commute:

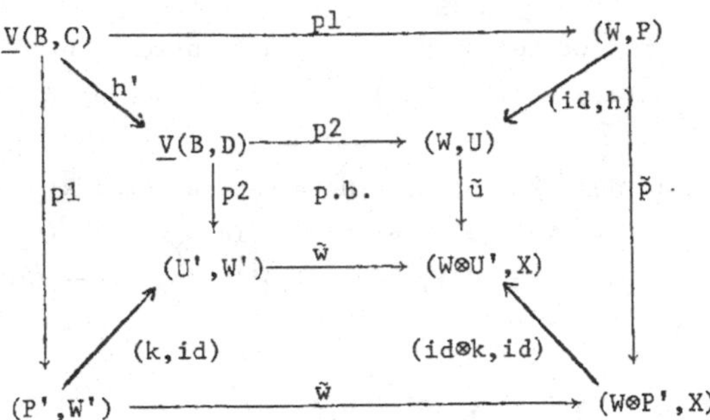

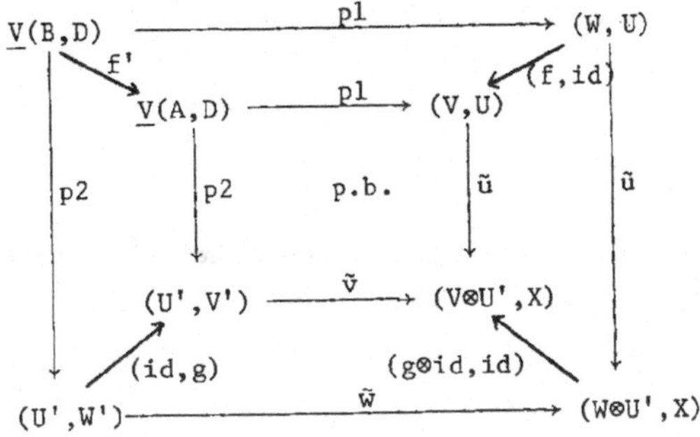

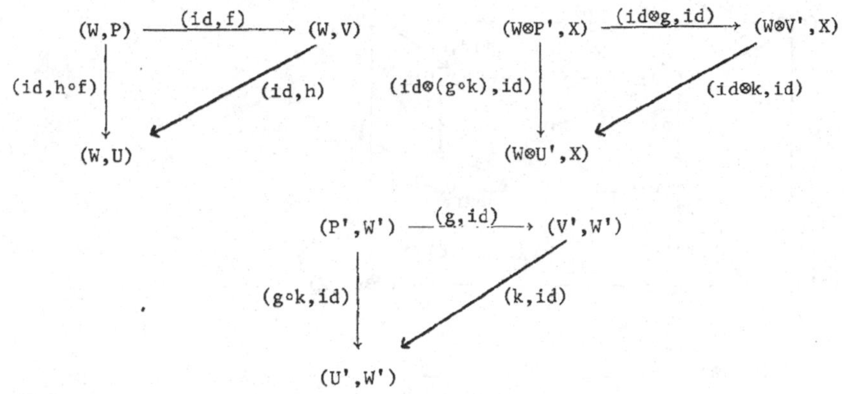
(W,P)
(id,f)
(W,V)
(id,h∘f)
(id,h)
(W,U)
(W⊗P',X)
(id⊗g,id)
(W⊗V',X)
(id⊗(g∘k),id)
(id⊗k,id)
(W⊗U',X)
(P',W')
(g,id)
(V',W')
(g∘k,id)
(k,id)
(U',W')

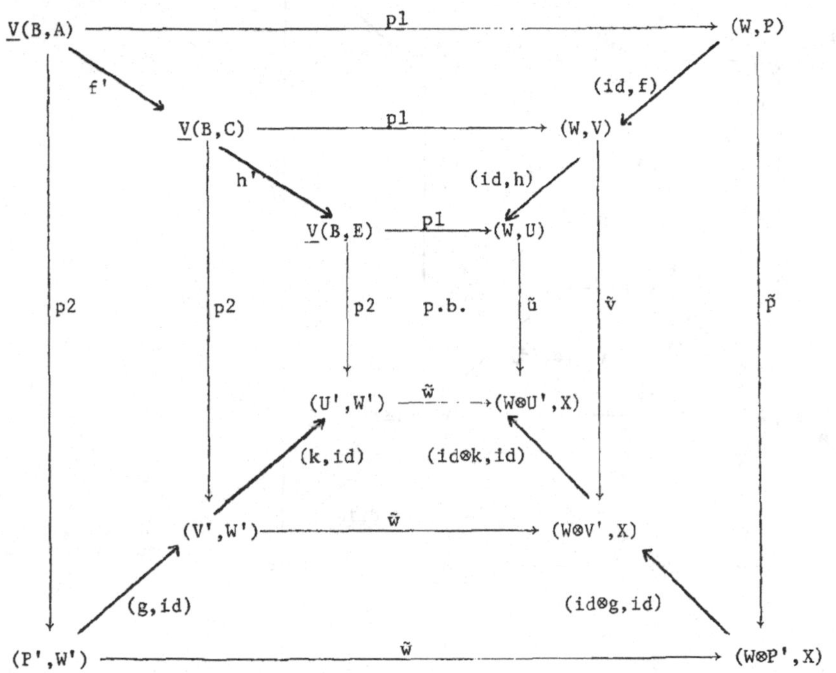
V(B,A)
p1
(W,P)
f'
(id,f)
V(B,C)
p1
(W,V)
h'
(id,h)
V(B,E)
p1
(W,U)
p2
p2
p2
p.b.
ũ
ṽ
p̃
(U',W')
w̃
(W⊗U',X)
(k,id)
(id⊗k,id)
(V',W')
w̃
(W⊗V',X)
(g,id)
(id⊗g,id)
(P',W')
w̃
(W⊗P',X)

FIGURE 3.

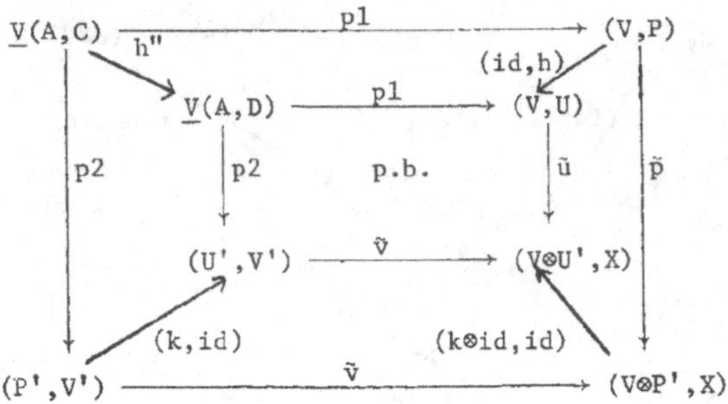

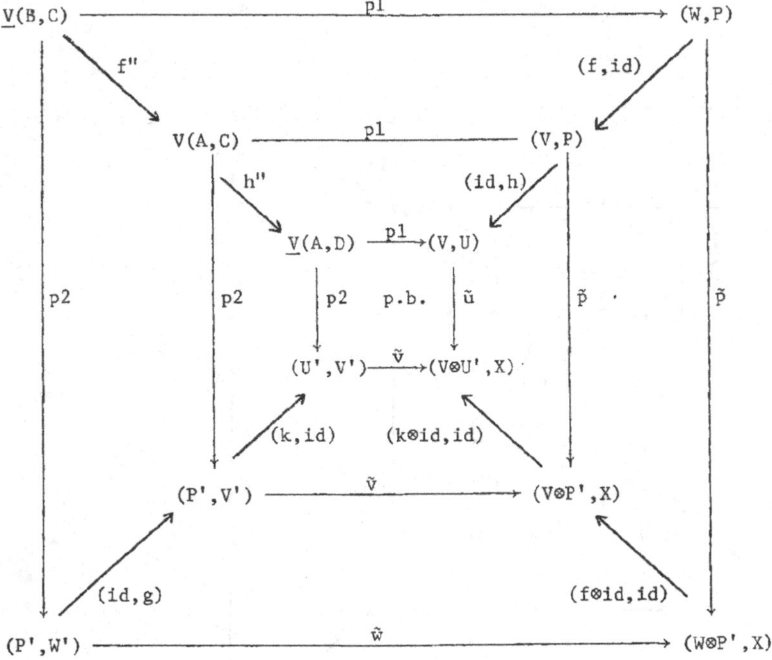

FIGURE 4.

This implies that the first diagram in Figure 4 commutes which implies, in turn, that the second one does.

Applying the same argument, the map from $\underline{V}(B,C)$ to $\underline{V}(A,D)$ induced by pullback is the same as $h''\circ f''$, hence the following diagram commutes:

$$\begin{array}{ccc} \underline{V}(B,C) & \xrightarrow{f''} & \underline{V}(A,C) \\ \downarrow{\scriptstyle h'} & & \downarrow{\scriptstyle h''} \\ \underline{V}(B,D) & \xrightarrow{f'} & \underline{V}(A,D) \end{array}$$

Next consider Figure 5. Since the center square of the first diagram is a pullback, $f' \circ h'$ is the unique map $\underline{V}(B,C) \longrightarrow \underline{V}(A,D)$ that makes the diagram commute.

Next consider the lower diagram of Figure 5. Using this and the fact that the following diagram commutes:

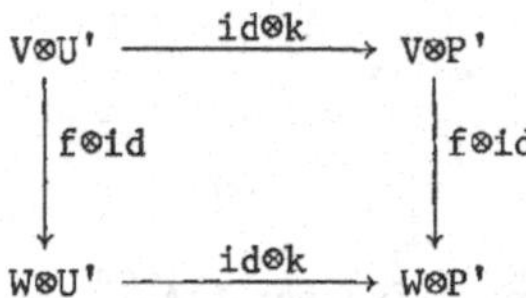

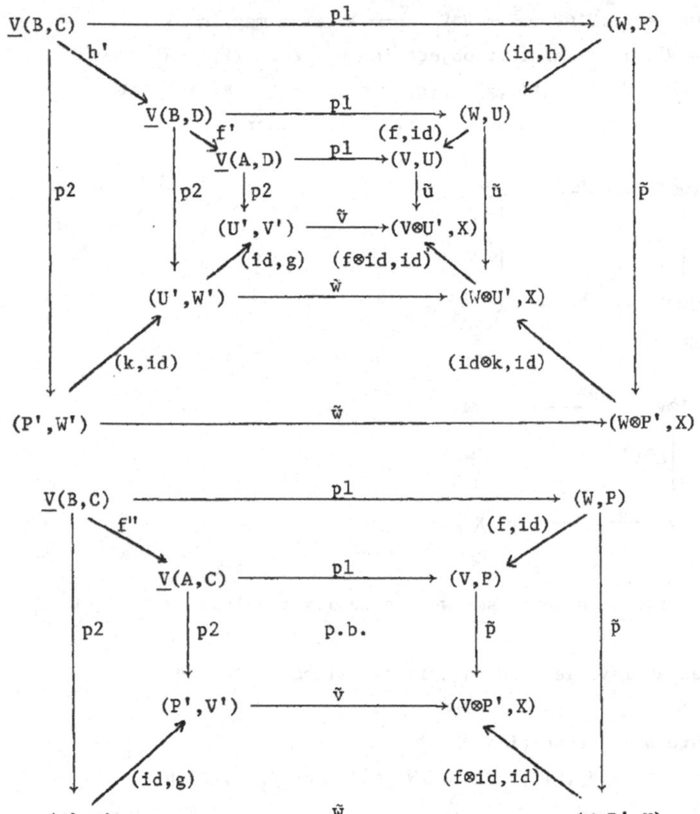

FIGURE 5.

we obtain the desired result that the diagram

commutes.

2. <u>The Functor *</u>.

In this section we shall define a functor $*:\underline{A}_X^{op} \longrightarrow \underline{A}_X$ and examine its relationship with $\underline{A}_X(-,-)$.

<u>Definition</u>. Given any object $A = (V,V',v)$ in $\underline{A}_X$ define $*(A)$ to be the object $(V',V,v\circ s)$ where $s:V'\otimes V \xrightarrow{s} V\otimes V' \xrightarrow{v} X$ is a map in $\underline{V}$.

Suppose $B = (W,W',w)$ is another object in $\underline{A}_X$ and $(f,g):A \longrightarrow B$ a morphism in $\underline{A}_X$, then define $*(f,g) = (g,f):*(B) \longrightarrow *(A)$. This definition is justified since the commutativity of the diagram:

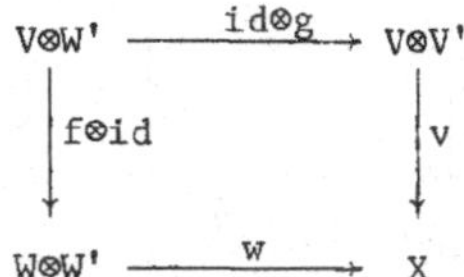

implies that the diagram

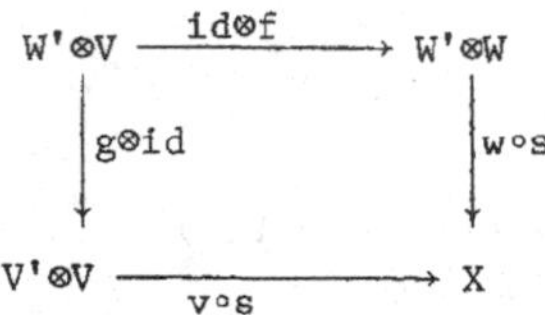

commutes.

From the above formula on morphisms we can easily conclude that $*$ is a functor.

Moreover $*$ has an inverse (contravariant), since $*\circ* = id_{\underline{A}_X}$.

The following are some properties of $*$:

<u>Proposition 1</u>. Given $A = (V,V',v)$, $B = (W,W',w')$ in $\underline{A}_X$, then $\underline{V}(A,B) \cong \underline{V}(*(B),*(A))$.

PROOF. By definition $*(A) = (V',V,v\circ s)$ and $*(B) = (W',W,w\circ s)$. Consider the commutative diagram of Figure 6.

Notice that the coherence of $\underline{V}$ implies squares (1), (2), (3), (4) commute. It also implies that the diagram

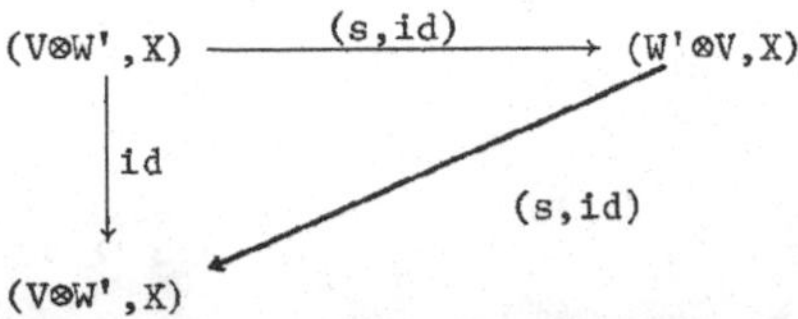

commutes.

The fact that

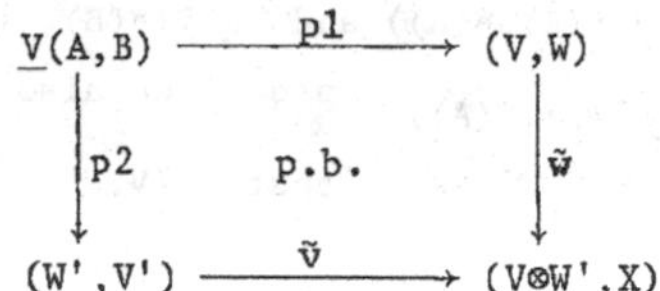

is a pullback square implies that there exists a unique $p:\underline{V}(A,B) \longrightarrow \underline{V}(*(B),*(A))$ such that the diagram of Figure 6 still commutes. Similarly the pullback square involve $\underline{V}(A,B)$ induces a unique map $q:\underline{V}(*(B),*(A)) \longrightarrow \underline{V}(A,B)$ such that the diagram of Figure 6 commutes. This implies $q \circ p$ is the map induced by the pullback square:

$$\begin{array}{ccc} \underline{V}(A,B) & \xrightarrow{\ p1\ } & (V,W) \\ \downarrow{\scriptstyle p2} & \text{p.b.} & \downarrow{\scriptstyle \tilde{w}} \\ (W',V') & \xrightarrow{\ \tilde{v}\ } & (V\otimes W',X) \end{array}$$

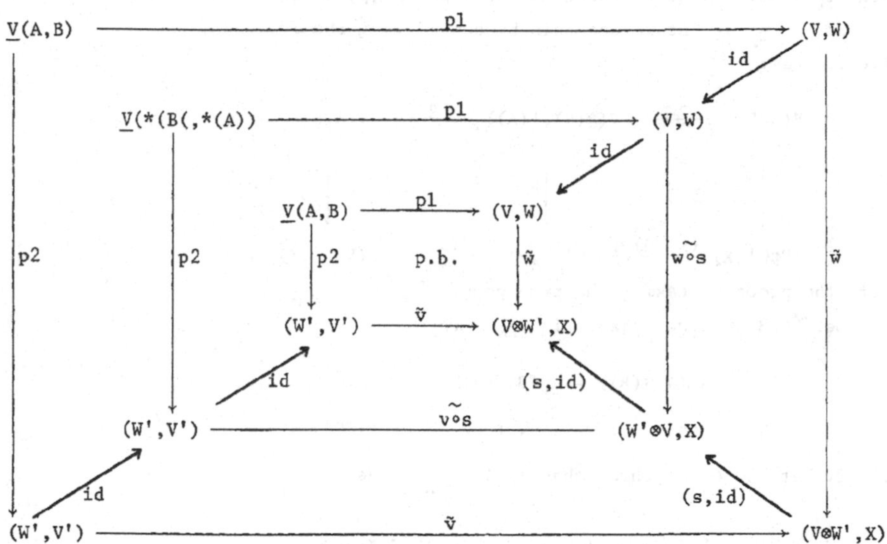

FIGURE 6.

But by the remark above $id_{\underline{V}(A,B)}$ also has this property. Hence $q \circ p = id_{\underline{V}(A,B)}$.

Now switch $\underline{V}(A,B)$ and $\underline{V}(*(B),*(A))$ in the previous diagram, and apply the same argument to conclude that $p \circ q = id_{\underline{V}(*(B),*(A))}$. This completes the proof.

Corollary. Let A,B be two objects in $\underline{A}_X$, then $\underline{V}(A,*(B)) = \underline{V}(B,*(A))$

PROOF. For any object C in $\underline{A}_X$, $*(*(C)) = C$.

Corollary. Let $A = (V,V',v)$, $B = (W,W',w)$, be two objects in $\underline{A}_X$, then $\underline{A}_X(A,B) \cong \underline{A}_X(*(B),*(A))$.

PROOF. By definition $*(A) = (V',V,v \circ s)$, $*(B) = (W',W,w \circ s)$ which implies that $\underline{A}_X(*(B),*(A)) = (\underline{V}(*(B),*(A)),W' \otimes V,n_1)$.

But recall that $\underline{A}_X(A,B) = (\underline{V}(A,B),V \otimes W',n_2)$; moreover we have isomorphism $p:\underline{V}(A,B) \longrightarrow \underline{V}(*(B),*(A))$ and $q:\underline{V}(*(B),*(A)) \longrightarrow \underline{V}(A,B)$ such that $id_{\underline{V}(A,B)} = q \circ p$, $id_{\underline{V}(*(B),*(A))} = p \circ q$. We also have $s(V,W'):V \otimes W' \longrightarrow W' \otimes V$ and $S(W',V):W' \otimes V \longrightarrow V \otimes W'$ such that $s(V,W') \circ s(W',V) = id_{W' \otimes V}$ and $s(W',V) \circ s(V,W') = id_{V \otimes W'}$.

Hence it is sufficient to check that the pair $(p,s(W',V))$ is indeed an isomorphism in $\underline{A}_X$. But we see this by considering the following commutative diagram:

$$\begin{array}{ccccc}
\underline{V}(A,B) & \xrightarrow{p} & \underline{V}(*(B),*(A)) & \xrightarrow{q} & \underline{V}(A,B) \\
\downarrow & & \downarrow & & \downarrow \\
(V \otimes W',X) & \xrightarrow{(s,id)} & (W' \otimes V,X) & \xrightarrow{(s,id)} & (V \otimes W',X)
\end{array}$$

and complete the proof by taking the transpose.

Corollary. Let A,B be two objects in $\underline{A}_X$, then

$$\underline{A}_X(A,*(B)) = \underline{A}_X(B,*(A)) .$$

PROOF. If C is an object in $\underline{A}_X$, then $C = *(*C))$.

Proposition 2. Let A,B,C be three objects in $\underline{A}_X$, then

$$\underline{V}(A,\underline{A}_X(B,*(C))) \cong \underline{V}(C,\underline{A}_X(B,*(A))).$$

PROOF. Let $A = (V,V',v)$, $B = (W,W',w)$, $C = (U,U',u)$. Then by definition $*(A) = (V',V,v \circ s)$ and $*(C) = (U',U,u \circ s)$.

Now put $Bc = \underline{A}_X(B,*(C)) = (\underline{V}(B,*(C)),W \otimes U,n_1)$ and

$$Ba = \underline{A}_X(B,*(A)) = (\underline{V}(B,*(A)),W \otimes V,n_2) .$$

Recall that $\underline{V}(B,*(C))$ and $\underline{V}(B,*(A))$ make the following squares pullbacks:

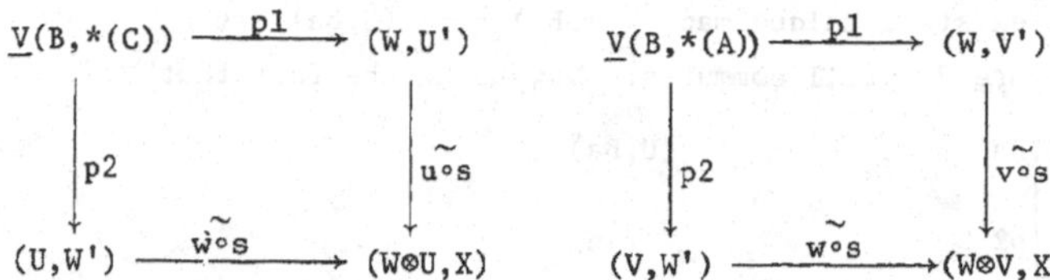

Now consider Figure 7. Since (U,-) and (V,-) have left adjoints, they preserve pullbacks, hence the outer and inner squares are still commutative.

But (1) is a pullback, hence the following diagrams commute:

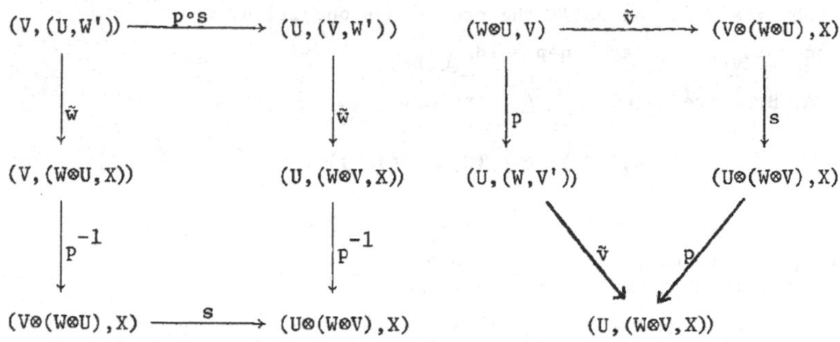

FIGURE 7.

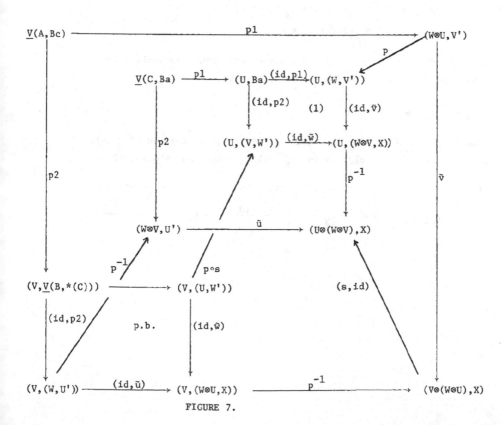

This implies that there exists a unique map $\underline{V}(A,Bc) \longrightarrow (U,Ba)$ such that the diagram of Figure 7 still commutes. Now using the fact that

$$\begin{array}{ccc} \underline{V}(C,Ba) & \xrightarrow{p1} & (U,Ba) \\ \downarrow{\scriptstyle p2} & & \downarrow{\scriptstyle \tilde{n}_2} \\ (W\otimes V,U') & \xrightarrow{\tilde{u}} & (U\otimes(W\otimes V),X) \end{array}$$

is a pullback, there exists a unique map $\underline{V}(A,Bc) \xrightarrow{q} \underline{V}(C,Ba)$.

A similar argument (Figure 8) shows the existence of a map $p:\underline{V}(C,Ba) \longrightarrow \underline{V}(A,Bc)$.

Applying the same argument as in the previous proposition, we conclude that $p\circ q = id_{\underline{V}(A,Bc)}$ and $q\circ p = id_{\underline{V}(C,Ba)}$.

Corollary. If A, B, C are objects in $\underline{A}_X$, then

$$\underline{A}_X(A,\underline{A}_X(B,*(C))) \cong \underline{A}_X(C,\underline{A}_X(B*(A))) \ .$$

PROOF. Apply the same argument as in previous corollary.

Corollary. Let A, B, C be objects in $\underline{A}_X$, then

$$\underline{A}_X(A,\underline{A}_X(B,C)) \cong \underline{A}_X(*(C),\underline{A}_X(B,*(A))) \ .$$

PROOF.
$$\begin{aligned} \underline{A}_X(A,\underline{A}_X(B,C)) &\cong \underline{A}_X(*(*(A)),\underline{A}_X(B,*(*(C)))) \ . \\ &\cong \underline{A}_X(*(C),\underline{A}_X(B,*(A))) \ . \end{aligned}$$

Remark. These propositions and corollaries concerning the duality lay the foundation of our construction, as we shall see later on.

3. The Functor $-\otimes-$

Note: Henceforth we write, for an object A of $\underline{A}_X$, A* instead of *(A).

Definition. Given A,B objects in $\underline{A}_X$, then define $A\otimes B = \underline{A}_X(A,B^*)^*$.

It is clear that $-\otimes-$ is a bifunctor, since $-\otimes-$ is the composition

$$\underline{A}_X \times \underline{A}_X \xrightarrow{(*,*)} \underline{A}_X^{\circ} \times \underline{A}_X^{\circ} \xrightarrow{(id,*)} \underline{A}_X^{\circ} \times \underline{A}_X \xrightarrow{\underline{A}_X(-,-)} \underline{A}_X$$

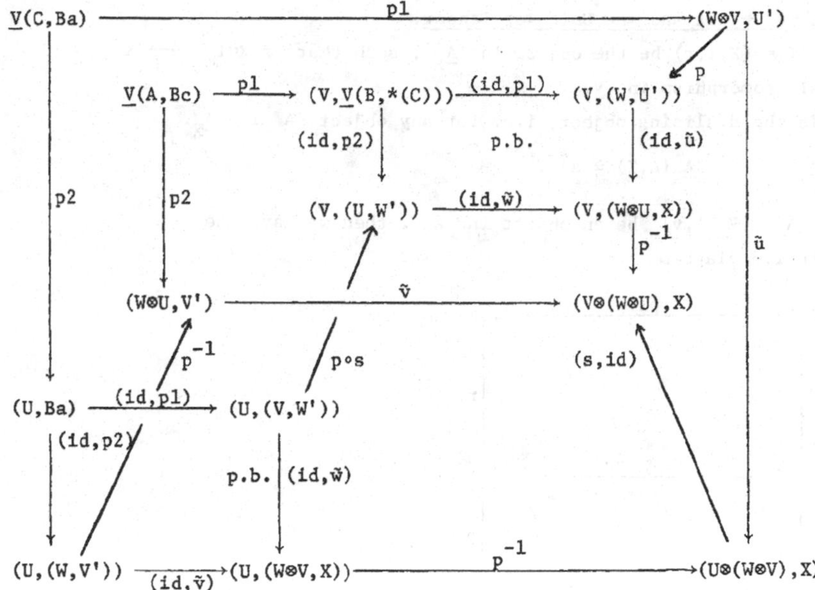

FIGURE 8.

<u>Proposition</u>. Let A,B be objects in $\underline{A}_X$, then

$$A \otimes B \cong B \otimes A.$$

PROOF. $$A \otimes B = \underline{A}_X(A,B^*)^* \cong \underline{A}_X(B,A^*)^* = B \otimes A.$$

<u>Proposition</u>. Let A,B,C be objects in $\underline{A}_X$, then

$$(A \otimes B)\ C \cong A \otimes (B \otimes C)$$

PROOF.
$$\begin{aligned}
(A \otimes B) \otimes C &= \underline{A}_X(A,B^*)^* \otimes C \\
&= \underline{A}_X(\underline{A}_X(A,B^*)^*,C^*)^* \\
&\cong \underline{A}_X(C,\underline{A}_X(A,B^*))^* \\
&\cong \underline{A}_X(C,\underline{A}_X(B,A^*))^* \\
&\cong \underline{A}_X(A,\underline{A}_X(B,C^*))^* \\
&\cong \underline{A}_X(\underline{A}_X(B,C^*)^*,A^*)^* \\
&\cong \underline{A}_X(A,\underline{A}_X(B,C^*)^{**})^* \\
&= A \otimes \underline{A}_X(B,C^*)^* \\
&= A \otimes (B \otimes C).
\end{aligned}$$

4. The Dualising Object and the Unit for Tensor.

Let $T = (X,I,r)$ be the object in $\underline{A}_X$, such that $r:X\otimes I \longrightarrow X$ is the cannonical isomorphism in $\underline{V}$.

Claim. T is the dualising object, i.e. for any object A in $\underline{A}_X$.

$$\underline{A}_X(A,T) \cong A^*.$$

PROOF. Let $A = (V,V',v)$ be an object in $\underline{A}_X$, then we have the following commutative diagram

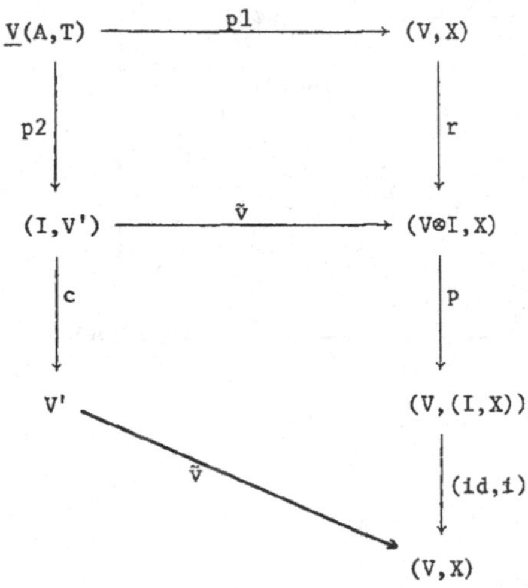

But

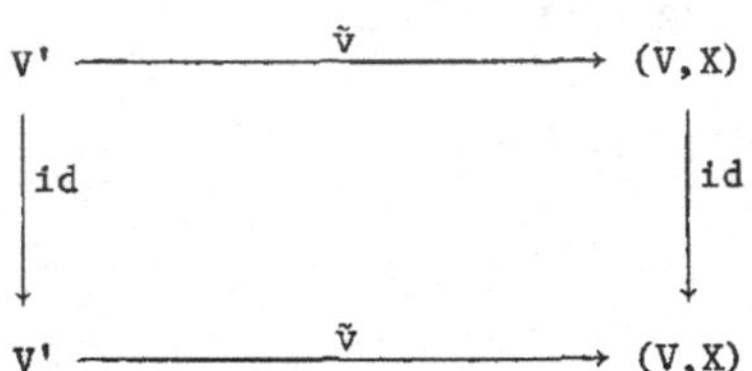

is trivially a pullback in $\underline{V}$, which implies that we have an induced (unique) morphism $f:\underline{V}(A,T) \longrightarrow V'$. Apply the same argument to get a unique map $g:V' \longrightarrow \underline{V}(A,T)$ such that $f\circ g = id_{V'}$ and $g\circ f = id_{\underline{V}(A,T)}$.

Corollary. T^* is the identity for $-\otimes-$.

PROOF. Suppose A is an object in $\underline{A}_X$, then

$$\begin{aligned} T^*\otimes A &= \underline{A}_X(T^*,A^*)^* \\ &\cong \underline{A}_X(A,T)^* \\ &\cong A^{**} \\ &= A \end{aligned}$$

On the other hand, $A\otimes T^* \cong T^*\otimes A \cong A$. This completes the proof.

Theorem. Let A,B,C be objects in $\underline{A}_X$, then

$$\underline{A}_X(A\otimes B,C) \cong \underline{A}_X(A,\underline{A}_X(B,C)) .$$

PROOF.

$$\begin{aligned}\underline{A}_X(A\otimes B,C) &= \underline{A}_X(\underline{A}_X(A,B^*)^*,C)\\ &\cong \underline{A}_X(C^*,\underline{A}_X(A,B^*))\\ &\cong \underline{A}_X(C^*,\underline{A}_X(B,A^*))\\ &\cong \underline{A}_X(A,\underline{A}_X(B,C)).\end{aligned}$$

Proposition. Let A be an object in $\underline{A}_X$ then

$$\underline{A}_X(T^*,A) \cong A.$$

PROOF.

$$\underline{A}_X(T^*,A) \cong \underline{A}_X(A^*,T) \cong A.$$

Remark.

(1) There is an obvious embedding functor from the comma category $(\underline{V},X)$ to $\underline{A}_X$ sending $V \longrightarrow X$ to $V\otimes I \longrightarrow X$: hence in this context $(\underline{V},X)$ has a $*$-autonomous structure.

(2) It is easy to verify $\underline{A}_X$ also satisfies our first assumption, i.e. the MacLane-Kelly coherence conditions.

CHAPTER IV: APPLICATIONS

1. Functor Categories

In this chapter, we shall apply the theory developed thus far to the double envelope of a symmetric monoidal category $\underline{C}$.

Before defining the double envelope, let us recall some elementary results of the functor categories.

Given categories $\underline{X}$ and $\underline{Y}$ we have the functor category $\underline{W} = \underline{X}^{\underline{Y}}$. We know that if $\underline{X}$ is complete, then so is $\underline{W}$, in the case $\underline{X} = \underline{S}$, the category of sets $\underline{W}$ also has a closed symmetric monoidal structure. The tensor is the cartesian product while the internal G^F is defined as the functor whose value at D is the set of nature transformations $F(-) \times \mathrm{Hom}(D,-) \longrightarrow G(-)$.

2. The Double Envelope.

Definition. Given a symmetric monoidal category with a faithful functor $|-|:\underline{C}^o \longrightarrow \underline{S}$, we denote the double envelope of $\underline{C}$ by $E(\underline{C})$. The objects of $E(\underline{C})$ are all triplets $(F,G;t)$ where F and G are functors from $\underline{C}^o$ to $\underline{S}$, t is a natural transformation from $F \times G$ to $|-\otimes-|$. A morphism from $(F,G;t)$ to $(F',G';s)$ in $E(\underline{C})$ is a pair (f,g) where f is a natural transformation from F to F' and g is a natural transformation from G' to G such that the following diagram

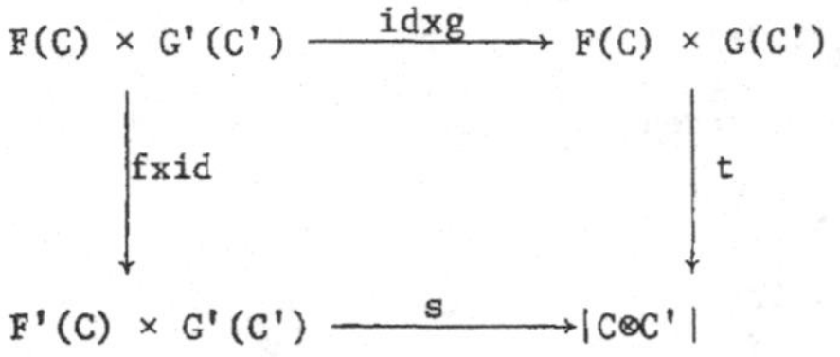

$$\begin{array}{ccc} F(C) \times G'(C') & \xrightarrow{id \times g} & F(C) \times G(C') \\ \Big\downarrow f \times id & & \Big\downarrow t \\ F'(C) \times G'(C') & \xrightarrow{s} & |C \otimes C'| \end{array}$$

commutes for every object (C,C') of $\underline{C}^o \times \underline{C}^o$.

Proposition. $E(\underline{C})$ is a category.

PROOF. Suppose $(f,g): (F,G;t) \longrightarrow (F',G';s)$

$(f',g'): (F',G';s) \longrightarrow (F'',G'';u)$ are maps

in $E(\underline{C})$, then the following diagram commutes for every (C,C') in $\underline{C}^o \times \underline{C}^o$.

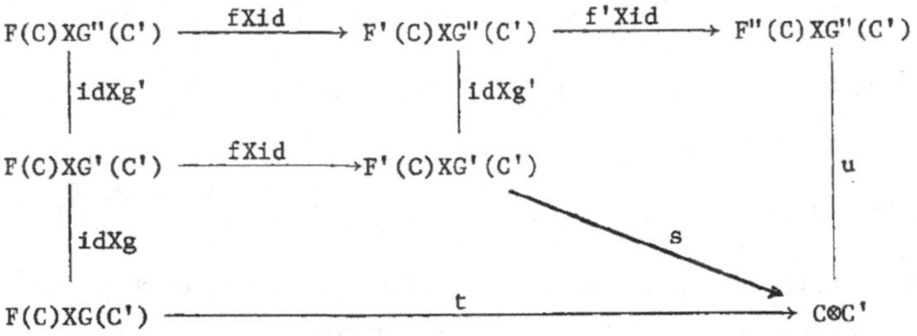

$$\begin{array}{ccccc} F(C) \times G''(C') & \xrightarrow{f \times id} & F'(C) \times G''(C') & \xrightarrow{f' \times id} & F''(C) \times G''(C') \\ \Big| id \times g' & & \Big| id \times g' & & \Big| \\ F(C) \times G'(C') & \xrightarrow{f \times id} & F'(C) \times G'(C') & & \Big| u \\ \Big| id \times g & & & \searrow s & \Big| \\ F(C) \times G(C') & & \xrightarrow{t} & & C \otimes C' \end{array}$$

This implies that $(f,g):(F,G;t) \longrightarrow (F',G';s)$

$(f',g'):(F',G';s) \longrightarrow (F'',G'';u)$

$(f'',g''):(F'',G'';u) \longrightarrow (F''',G''';v)$

are maps in $E(\underline{C})$, then

$$\begin{aligned} (f'',G'')\circ((f',g')\circ(f,g)) &= (f'',g'')\circ(f'\circ f,g\circ g') \\ &= (f''\circ(f'\circ f),(g\circ g')\circ g'') \\ &= ((f'\circ f')\circ f,g\circ(g'\circ g'')) \\ &= (f''\circ f',g'\circ g'')\circ(f,g) \\ &= ((f'',g'')\circ(f',g'))\circ(f,g). \end{aligned}$$

Moreover, given $(F,g;t)$ then (id_F,id_G) is the obvious choice for identity.

Before proving the main theorem of this chapter, let us investigate the functor categories $\underline{S}^{\underline{C}^o}$ and $\underline{S}^{\underline{C}^o \times \underline{C}^o}$. There are two obvious embeddings of $\underline{S}^{\underline{C}^o}$ into $\underline{S}^{\underline{C}^o \times \underline{C}^o}$, namely ℓ and r, where $\ell(F) = F \times I$ and $r(F) = I \times F$ for every F in $\underline{S}^{\underline{C}^o}$, and I is the unit in $\underline{S}^{\underline{C}^o}$ i.e. I sends every object into the singleton (the terminal object) in S. Hence we can regard objects in $\underline{S}^{\underline{C}^o}$ as objects in $\underline{S}^{\underline{C}^o \times \underline{C}^o}$ via either embedding. Now we can prove.

Proposition. $E(\underline{C})$ is enriched over $\underline{V} = \underline{S}^{\underline{C}^o \times \underline{C}^o}$.

PROOF. By previous remark $\underline{V}$ is a closed symmetric monoidal category with pullbacks, moreover it is coherent.

Now given $A = (G,F;t)$ and $B = (G',F';s)$ in $E(\underline{C})$ we have to define $\underline{V}(A,B)$ an object in $\underline{V}(= \underline{S}^{\underline{C}^o \times \underline{C}^o})$. Suppose (C,C') is an object of $\underline{S}^{\underline{C}^o \times \underline{C}^o}$, then $\underline{V}(A,B)$ is the functor whose value at (C,C') is defined by requiring that the diagram

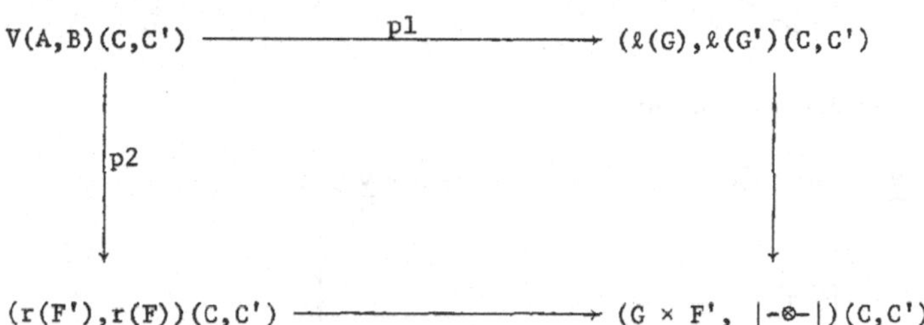

be a pullback.

Note. $(-,-)$ denotes the internal hom-functor of $\underline{S}^{\underline{C}^o \times \underline{C}^o}$. As for the map $(\ell(G),\ell(G'))(C,C') \longrightarrow (G \times F', |-\otimes-|)(C,C')$, we simply observe that in $\underline{V}$, $G \times F'$ is isomorphic to $\ell(G) \times r(F')$. Then the adjoint property of $\underline{V}$ constructs such a map (in the same fashion as in Chapter II, Section 2.) A similar argument constructs map $(r(F'),r(F))(C,C') \longrightarrow (G \times F, |-\otimes-|)(C,C')$.

Now the enrichment follows immediately from the result in Chapter II, since this is how pullbacks are defined in the functor category, i.e. by point-wise evaluation. This concludes the proof.

Theorem. $E(\underline{C})$ is a subcategory of a *-autonomous category $\underline{A}$; moreover $\underline{A}$ is enriched over $\underline{V}$.

PROOF. Put $X = |-\otimes-|$, then follow the construction in Chapter III.

3. Miscellaneous Results.

In this section, we are assuming $\underline{V}$ has all the properties as given in Chapter I and we shall prove that there is a functor F maps $\underline{V}$ to $\underline{V}$ - CAT($\underline{V}$ - CAT is the category of all categories which are enriched over $\underline{V}$).

The functor F on objects of $\underline{V}$ is obvious: given X in $\underline{V}$, then put $F(X) = \underline{A}_X$.

Now we have to show given a map $f:X \longrightarrow S$ in $\underline{V}$, this induces a $\underline{V}$-functor $T(= F(f))$ from $\underline{A}_X$ to $\underline{A}_S$.

The notion of a $\underline{V}$-functor can be found in [Eilenberg & Kelly] Chapter II, Section 6. In this case we have to show:

(i) a function T maps objects of $\underline{A}_X$ to objects of $\underline{A}_S$.

(ii) for each B,C in $\underline{A}_X$, a morphism $T(B,C)$ maps $\underline{V}(B,C)$ to $\underline{V}(T(B),T(C))$ in $\underline{V}$ such that the following axioms are satisfied:

(1) The following diagram commutes:

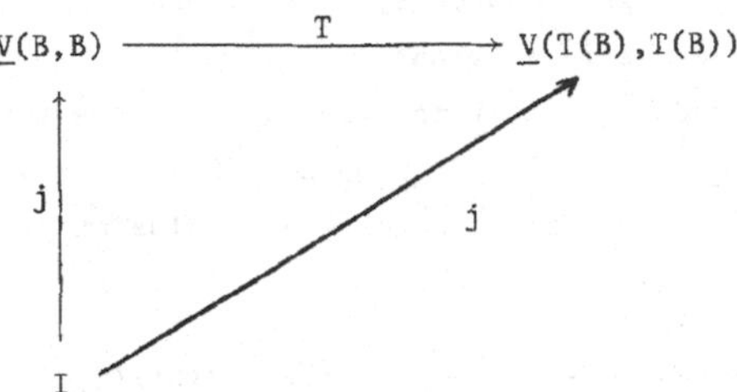

(2) The following diagram commutes:

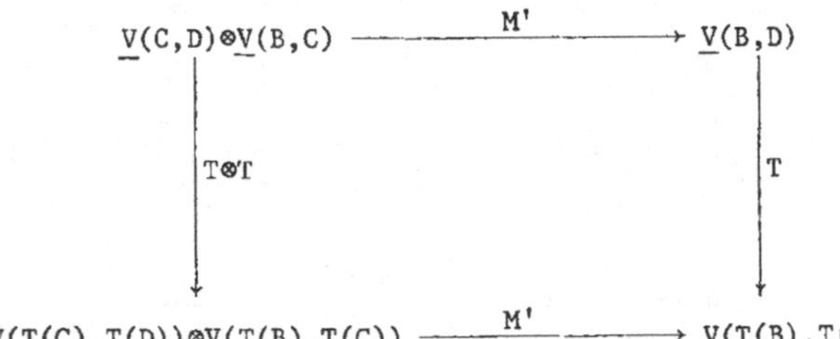

Note. In both categories we denote the enriched object by $\underline{V}(-,-)$, it is clear from the context which one we are referring to.

The function T on objects of $\underline{A}_X$ is obvious; given $B = (V,V',v)$ in $\underline{A}_X$, then $T(B)$ is the composition $V\otimes V' \xrightarrow{v} X \xrightarrow{f} S$ i.e. $T(B) = (V,V',f\circ v)$.

To show (ii):

Suppose $B = (V,V',v)$ $C = (W,W',w)$ objects in $\underline{A}_X$, then $T(B) = (V,V',f\circ v)$, $T(C) = (W,W',f\circ w)$ and the following diagram commutes

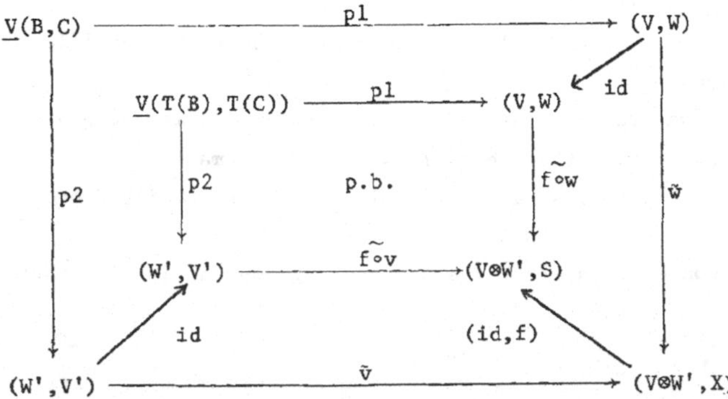

Since the inner square is a pullback, there exists a (unique) map $T(B,C)$ from $\underline{V}(B,C)$ to $\underline{V}(T(B),T(C))$.

To show (1) commutes let $B = (V,V',v)$ in $\underline{A}_X$. Then $T(B) = (V,V',f\circ v)$ and the following diagrams commute:

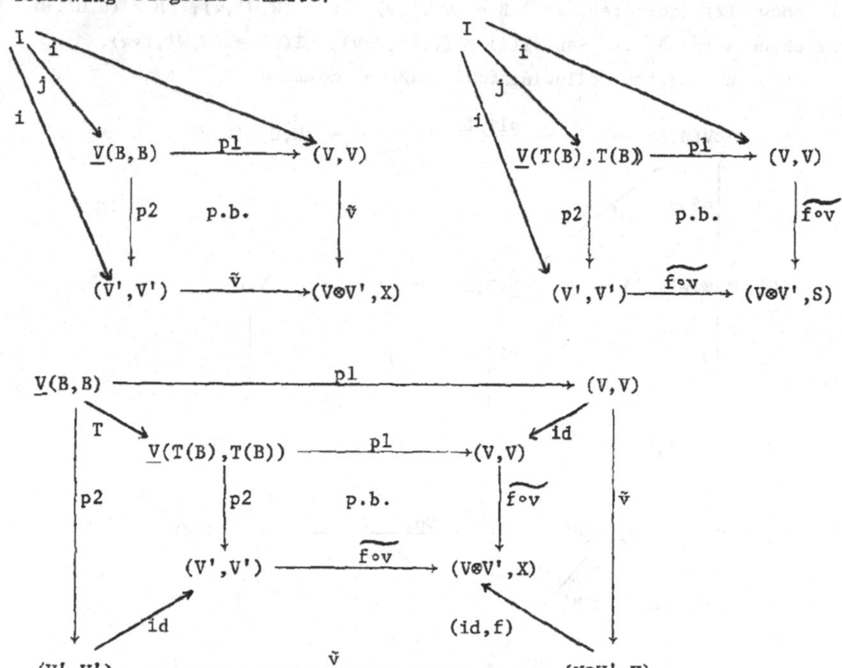

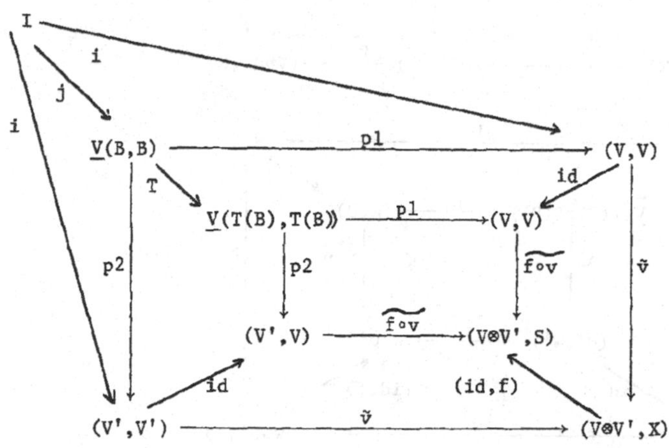

Hence the composition $I \xrightarrow{j} \underline{V}(B,B) \xrightarrow{T} \underline{V}(T(B),T(B))$ and map $I \xrightarrow{j} \underline{V}(T(B),T(B))$ are both induced by pulling back. Thus by the uniqueness property they are "equal".

To show (2) commutes, let $B = (V,V',v)$, $C = (W,W',w)$, $D = (U,U',u)$ be three objects in $\underline{A}_X$. Then $T(B) = (V,V',f\circ v)$, $T(C) = (W,W',f\circ w)$, $T(D) = (U,U',f\circ u)$ and the following four diagrams commute:

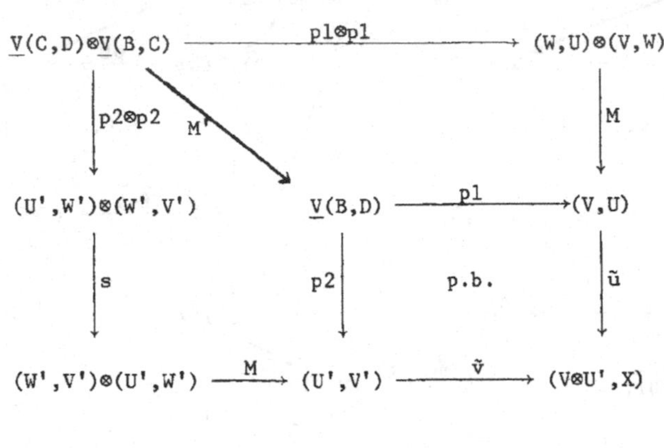

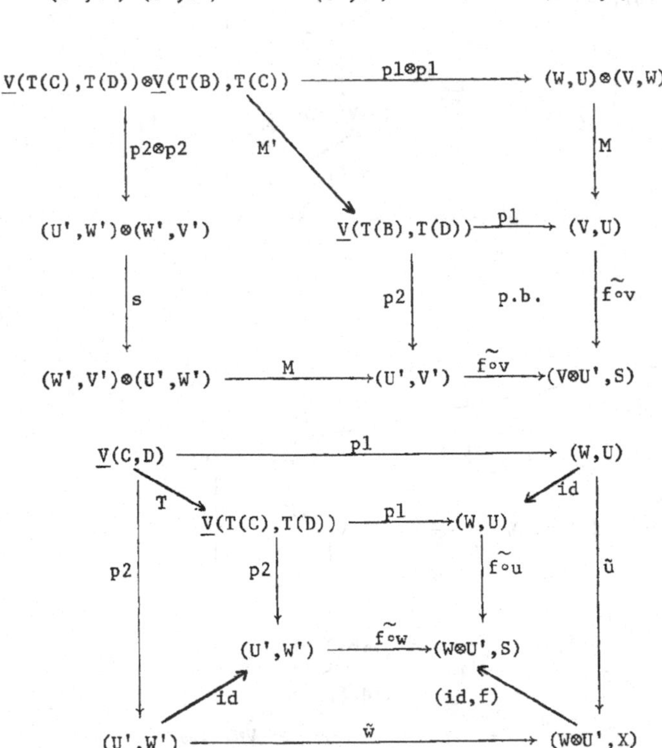

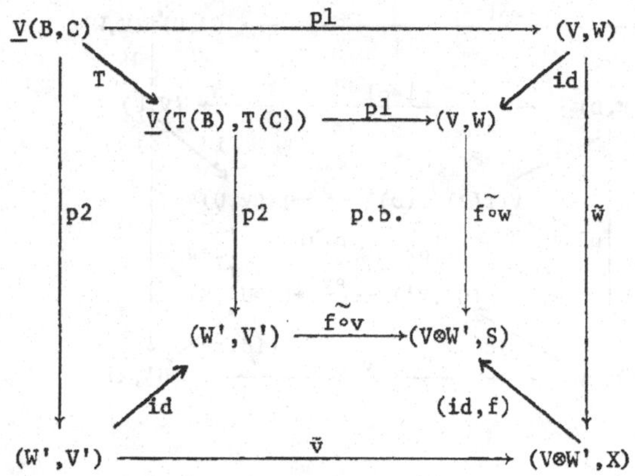
V(B,C)
p1
(V,W)
T
id
V(T(B),T(C))
p1
(V,W)
p2
p2
p.b.
f̃∘w
w̃
(W',V')
f̃∘v
(V⊗W',S)
id
(id,f)
(W',V')
ṽ
(V⊗W',X)

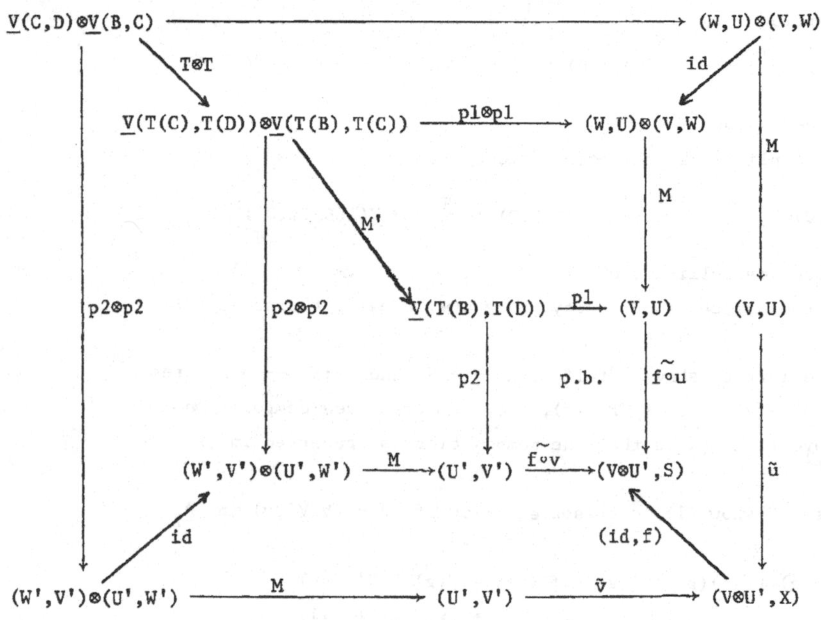
V(C,D)⊗V(B,C)
(W,U)⊗(V,W)
T⊗T
id
V(T(C),T(D))⊗V(T(B),T(C))
p1⊗p1
(W,U)⊗(V,W)
M
M
M'
p2⊗p2
p2⊗p2
V(T(B),T(D))
p1
(V,U)
(V,U)
p2
p.b.
f̃∘u
(W',V')⊗(U',W')
M
(U',V')
f̃∘v
(V⊗U',S)
ũ
id
(id,f)
(W',V')⊗(U',W')
M
(U',V')
ṽ
(V⊗U',X)

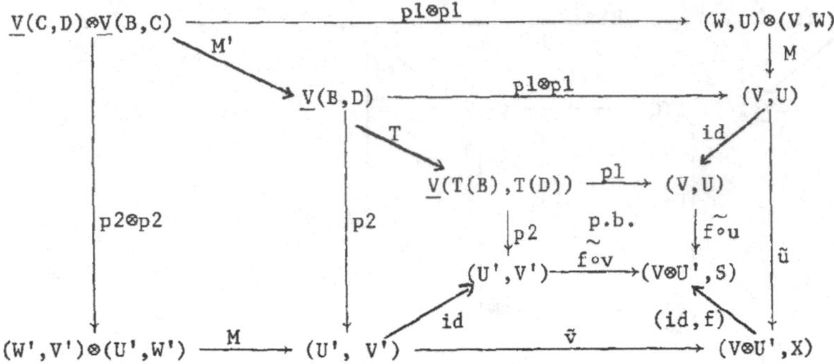

This implies that the diagrams above commute, which implies that the composition

$$\underline{V}(C,D)\otimes(\underline{V}(B,C)) \xrightarrow{T\otimes T} \underline{V}(T(C),T(D))\otimes\underline{V}(T(B),T(C)) \xrightarrow{M'} \underline{V}(T(B),T(D))$$

is the map induced by pulling back.

This also implies that the composition

$$\underline{V}(C,D)\otimes \underline{V}(B,C) \xrightarrow{M'} \underline{V}(B,D) \xrightarrow{T} \underline{V}(T(B),T(D))$$

is the map induced by pulling back.

Hence by the uniqueness property, they are "equal", therefore (2) commutes.

Now we are left to show that if $f:X \longrightarrow S$ and $g:S \longrightarrow K$ are maps in $\underline{V}$, then $F(g)\circ F(f) = F(g\circ f)$, i.e. F preserves composition.

All we have to check is that the composition is preserved in (i) and (ii).

It is easy to show (i) is preserved. For if $B = (V,V',v)$ in $\underline{A}_X$, then

$$\begin{aligned}(F(g)\circ F(f))(B) = F(g)(F(f)(B)) &= F(g)(V,V',f\circ v)\\ &= (V,V',g\circ(f\circ v))\\ &= (V,V',(g\circ f)\circ v)\\ &= F(g\circ f)(B).\end{aligned}$$

To show (ii) is preserved: Let $B = (V,V',v)$, $C = (W,W',w)$ in $\underline{A}_X$, then

$F(f)(B) = (V,V',f\circ v), F(f)(C) = (W,W',f\circ w), (F(g)\circ F(f))(B) = F(g\circ f)(B)$
$= (V,V',(g\circ f)\circ v), F(g\circ f)(C) = (F(g)\circ F(f))(C) = (W,W',(g\circ f)\circ w)$

and the diagrams (*), (**) and (***) commute

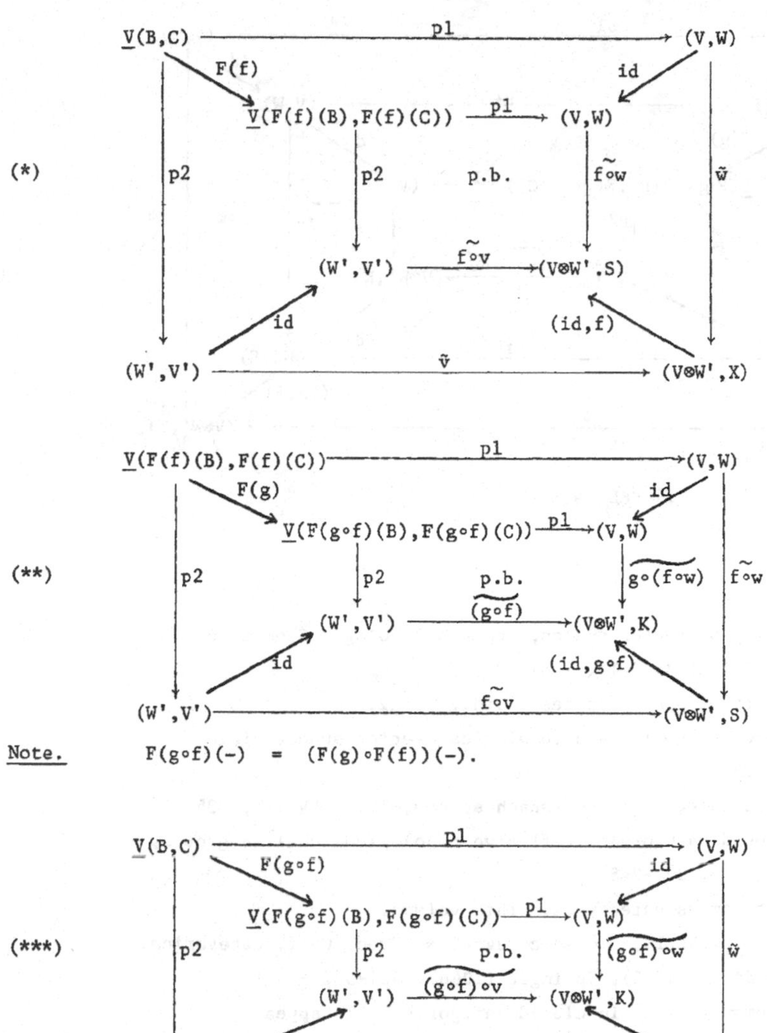

Note. $F(g\circ f)(-) = (F(g)\circ F(f))(-)$.

But (*) and (**) imply the diagram of Figure 9 commutes. This implies that both $F(g\circ f)$ in (***) and the composition

$$\underline{V}(B,C) \xrightarrow{F(f)} \underline{V}(F(f)(B),F(f)(C)) \xrightarrow{F(g)} \underline{V}(F(g\circ f)(B),F(g\circ f)(C))$$

are induced by pulling back. Hence it follows they are equal.

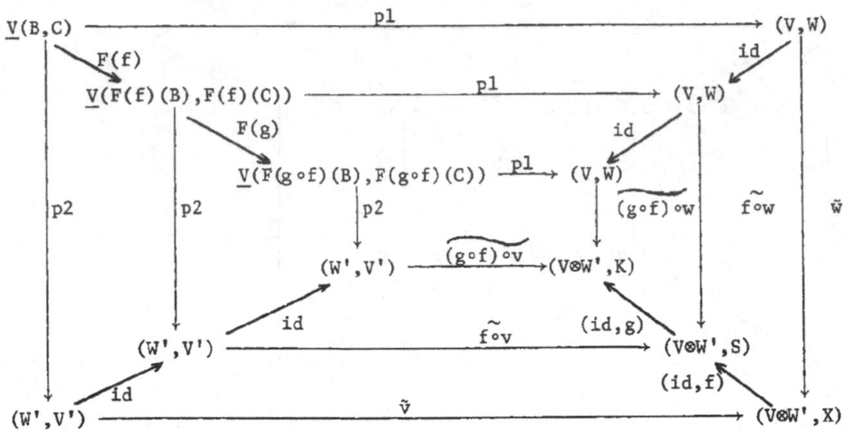

FIGURE 9.

BIBLIOGRAPHY

1. M. BARR, Duality of vector spaces, Cahiers Topologie Géométrie Différentielle, XVII-1 (1976), 3-14.
2. M. BARR, Duality of Banach spaces, Ibid., 15-32.
3. M. BARR, Closed categories and topological vector spaces, Ibid. XVII-3, 223-234.
4. M. BARR, Closed categories and Banach spaces, Ibid. XVII-4, 335-342.
5. M. BARR, A closed category of reflexive topological abelian groups, Ibid. XVIII-3, 221-248.
6. M. BARR, *-autonomous categories, This volume.
7. S. EILENBERG, G.M. KELLY, Closed categories, Proc. Conf. Categorical Alg. (La Jolla, 1965), Springer (1966), 421-562.
8. M.E. SZABO, Commutativity in closed categories. To appear.

Index of Definitions

Index of Notation

Vol. 580: C. Castaing and M. Valadier, Convex Analysis and Measurable Multifunctions. VIII, 278 pages. 1977.

Vol. 581: Séminaire de Probabilités XI, Université de Strasbourg. Proceedings 1975/1976. Edité par C. Dellacherie, P. A. Meyer et M. Weil. VI, 574 pages. 1977.

Vol. 582: J. M. G. Fell, Induced Representations and Banach *-Algebraic Bundles. IV, 349 pages. 1977.

Vol. 583: W. Hirsch, C. C. Pugh and M. Shub, Invariant Manifolds. IV, 149 pages. 1977.

Vol. 584: C. Brezinski, Accélération de la Convergence en Analyse Numérique. IV, 313 pages. 1977.

Vol. 585: T. A. Springer, Invariant Theory. VI, 112 pages. 1977.

Vol. 586: Séminaire d'Algèbre Paul Dubreil, Paris 1975–1976 (29ème Année). Edited by M. P. Malliavin. VI, 188 pages. 1977.

Vol. 587: Non-Commutative Harmonic Analysis. Proceedings 1976. Edited by J. Carmona and M. Vergne. IV, 240 pages. 1977.

Vol. 588: P. Molino, Théorie des G-Structures: Le Problème d'Equivalence. VI, 163 pages. 1977.

Vol. 589: Cohomologie l-adique et Fonctions L. Séminaire de Géométrie Algébrique du Bois-Marie 1965–66, SGA 5. Edité par L. Illusie. XII, 484 pages. 1977.

Vol. 590: H. Matsumoto, Analyse Harmonique dans les Systèmes de Tits Bornologiques de Type Affine. IV, 219 pages. 1977.

Vol. 591: G. A. Anderson, Surgery with Coefficients. VIII, 157 pages. 1977.

Vol. 592: D. Voigt, Induzierte Darstellungen in der Theorie der endlichen, algebraischen Gruppen. V, 413 Seiten. 1977.

Vol. 593: K. Barbey and H. König, Abstract Analytic Function Theory and Hardy Algebras. VIII, 260 pages. 1977.

Vol. 594: Singular Perturbations and Boundary Layer Theory, Lyon 1976. Edited by C. M. Brauner, B. Gay, and J. Mathieu. VIII, 539 pages. 1977.

Vol. 595: W. Hazod, Stetige Faltungshalbgruppen von Wahrscheinlichkeitsmaßen und erzeugende Distributionen. XIII, 157 Seiten. 1977.

Vol. 596: K. Deimling, Ordinary Differential Equations in Banach Spaces. VI, 137 pages. 1977.

Vol. 597: Geometry and Topology, Rio de Janeiro, July 1976. Proceedings. Edited by J. Palis and M. do Carmo. VI, 866 pages. 1977.

Vol. 598: J. Hoffmann-Jørgensen, T. M. Liggett et J. Neveu, Ecole d'Eté de Probabilités de Saint-Flour VI – 1976. Edité par P.-L. Hennequin. XII, 447 pages. 1977.

Vol. 599: Complex Analysis, Kentucky 1976. Proceedings. Edited by J. D. Buckholtz and T. J. Suffridge. X, 159 pages. 1977.

Vol. 600: W. Stoll, Value Distribution on Parabolic Spaces. VIII, 216 pages. 1977.

Vol. 601: Modular Functions of one Variable V, Bonn 1976. Proceedings. Edited by J.-P. Serre and D. B. Zagier. VI, 294 pages. 1977.

Vol. 602: J. P. Brezin, Harmonic Analysis on Compact Solvmanifolds. VIII, 179 pages. 1977.

Vol. 603: B. Moishezon, Complex Surfaces and Connected Sums of Complex Projective Planes. IV, 234 pages. 1977.

Vol. 604: Banach Spaces of Analytic Functions, Kent, Ohio 1976. Proceedings. Edited by J. Baker, C. Cleaver and Joseph Diestel. VI, 141 pages. 1977.

Vol. 605: Sario et al., Classification Theory of Riemannian Manifolds. XX, 498 pages. 1977.

Vol. 606: Mathematical Aspects of Finite Element Methods. Proceedings 1975. Edited by I. Galligani and E. Magenes. VI, 362 pages. 1977.

Vol. 607: M. Métivier, Reelle und Vektorwertige Quasimartingale und die Theorie der Stochastischen Integration. X, 310 Seiten. 1977.

Vol. 608: Bigard et al., Groupes et Anneaux Réticulés. XIV, 334 pages. 1977.

Vol. 609: General Topology and Its Relations to Modern Analysis and Algebra IV. Proceedings 1976. Edited by J. Novák. XVIII, 225 pages. 1977.

Vol. 610: G. Jensen, Higher Order Contact of Submanifolds of Homogeneous Spaces. XII, 154 pages. 1977.

Vol. 611: M. Makkai and G. E. Reyes, First Order Categorical Logic. VIII, 301 pages. 1977.

Vol. 612: E. M. Kleinberg, Infinitary Combinatorics and the Axiom of Determinateness. VIII, 150 pages. 1977.

Vol. 613: E. Behrends et al., L^p-Structure in Real Banach Spaces. X, 108 pages. 1977.

Vol. 614: H. Yanagihara, Theory of Hopf Algebras Attached to Group Schemes. VIII, 308 pages. 1977.

Vol. 615: Turbulence Seminar, Proceedings 1976/77. Edited by P. Bernard and T. Ratiu. VI, 155 pages. 1977.

Vol. 616: Abelian Group Theory, 2nd New Mexico State University Conference, 1976. Proceedings. Edited by D. Arnold, R. Hunter and E. Walker. X, 423 pages. 1977.

Vol. 617: K. J. Devlin, The Axiom of Constructibility: A Guide for the Mathematician. VIII, 96 pages. 1977.

Vol. 618: I. I. Hirschman, Jr. and D. E. Hughes, Extreme Eigen Values of Toeplitz Operators. VI, 145 pages. 1977.

Vol. 619: Set Theory and Hierarchy Theory V, Bierutowice 1976. Edited by A. Lachlan, M. Srebrny, and A. Zarach. VIII, 358 pages. 1977.

Vol. 620: H. Popp, Moduli Theory and Classification Theory of Algebraic Varieties. VIII, 189 pages. 1977.

Vol. 621: Kauffman et al., The Deficiency Index Problem. VI, 112 pages. 1977.

Vol. 622: Combinatorial Mathematics V, Melbourne 1976. Proceedings. Edited by C. Little. VIII, 213 pages. 1977.

Vol. 623: I. Erdelyi and R. Lange, Spectral Decompositions on Banach Spaces. VIII, 122 pages. 1977.

Vol. 624: Y. Guivarc'h et al., Marches Aléatoires sur les Groupes de Lie. VIII, 292 pages. 1977.

Vol. 625: J. P. Alexander et al., Odd Order Group Actions and Witt Classification of Innerproducts. IV, 202 pages. 1977.

Vol. 626: Number Theory Day, New York 1976. Proceedings. Edited by M. B. Nathanson. VI, 241 pages. 1977.

Vol. 627: Modular Functions of One Variable VI, Bonn 1976. Proceedings. Edited by J.-P. Serre and D. B. Zagier. VI, 339 pages. 1977.

Vol. 628: H. J. Baues, Obstruction Theory on the Homotopy Classification of Maps. XII, 387 pages. 1977.

Vol. 629: W.A. Coppel, Dichotomies in Stability Theory. VI, 98 pages. 1978.

Vol. 630: Numerical Analysis, Proceedings, Biennial Conference, Dundee 1977. Edited by G. A. Watson. XII, 199 pages. 1978.

Vol. 631: Numerical Treatment of Differential Equations. Proceedings 1976. Edited by R. Bulirsch, R. D. Grigorieff, and J. Schröder. X, 219 pages. 1978.

Vol. 632: J.-F. Boutot, Schéma de Picard Local. X, 165 pages. 1978.

Vol. 633: N. R. Coleff and M. E. Herrera, Les Courants Résiduels Associés à une Forme Méromorphe. X, 211 pages. 1978.

Vol. 634: H. Kurke et al., Die Approximationseigenschaft lokaler Ringe. IV, 204 Seiten. 1978.

Vol. 635: T. Y. Lam, Serre's Conjecture. XVI, 227 pages. 1978.

Vol. 636: Journées de Statistique des Processus Stochastiques, Grenoble 1977, Proceedings. Edité par Didier Dacunha-Castelle et Bernard Van Cutsem. VII, 202 pages. 1978.

Vol. 637: W. B. Jurkat, Meromorphe Differentialgleichungen. VII, 194 Seiten. 1978.

Vol. 638: P. Shanahan, The Atiyah-Singer Index Theorem, An Introduction. V, 224 pages. 1978.

Vol. 639: N. Adasch et al., Topological Vector Spaces. V, 125 pages. 1978.

Vol. 640: J. L. Dupont, Curvature and Characteristic Classes. X, 175 pages. 1978.

Vol. 641: Séminaire d'Algèbre Paul Dubreil, Proceedings Paris 1976–1977. Edité par M. P. Malliavin. IV, 367 pages. 1978.

Vol. 642: Theory and Applications of Graphs, Proceedings, Michigan 1976. Edited by Y. Alavi and D. R. Lick. XIV, 635 pages. 1978.

Vol. 643: M. Davis, Multiaxial Actions on Manifolds. VI, 141 pages. 1978.

Vol. 644: Vector Space Measures and Applications I, Proceedings 1977. Edited by R. M. Aron and S. Dineen. VIII, 451 pages. 1978.

Vol. 645: Vector Space Measures and Applications II, Proceedings 1977. Edited by R. M. Aron and S. Dineen. VIII, 218 pages. 1978.

Vol. 646: O. Tammi, Extremum Problems for Bounded Univalent Functions. VIII, 313 pages. 1978.

Vol. 647: L. J. Ratliff, Jr., Chain Conjectures in Ring Theory. VIII, 133 pages. 1978.

Vol. 648: Nonlinear Partial Differential Equations and Applications, Proceedings, Indiana 1976–1977. Edited by J. M. Chadam. VI, 206 pages. 1978.

Vol. 649: Séminaire de Probabilités XII, Proceedings, Strasbourg, 1976–1977. Edité par C. Dellacherie, P. A. Meyer et M. Weil. VIII, 805 pages. 1978.

Vol. 650: C*-Algebras and Applications to Physics. Proceedings 1977. Edited by H. Araki and R. V. Kadison. V, 192 pages. 1978.

Vol. 651: P. W. Michor, Functors and Categories of Banach Spaces. VI, 99 pages. 1978.

Vol. 652: Differential Topology, Foliations and Gelfand-Fuks-Cohomology, Proceedings 1976. Edited by P. A. Schweitzer. XIV, 252 pages. 1978.

Vol. 653: Locally Interacting Systems and Their Application in Biology. Proceedings, 1976. Edited by R. L. Dobrushin, V. I. Kryukov and A. L. Toom. XI, 202 pages. 1978.

Vol. 654: J. P. Buhler, Icosahedral Golois Representations. III, 143 pages. 1978.

Vol. 655: R. Baeza, Quadratic Forms Over Semilocal Rings. VI, 199 pages. 1978.

Vol. 656: Probability Theory on Vector Spaces. Proceedings, 1977. Edited by A. Weron. VIII, 274 pages. 1978.

Vol. 657: Geometric Applications of Homotopy Theory I, Proceedings 1977. Edited by M. G. Barratt and M. E. Mahowald. VIII, 459 pages. 1978.

Vol. 658: Geometric Applications of Homotopy Theory II, Proceedings 1977. Edited by M. G. Barratt and M. E. Mahowald. VIII, 487 pages. 1978.

Vol. 659: Bruckner, Differentiation of Real Functions. X, 247 pages. 1978.

Vol. 660: Equations aux Dérivée Partielles. Proceedings, 1977. Edité par Pham The Lai. VI, 216 pages. 1978.

Vol. 661: P. T. Johnstone, R. Paré, R. D. Rosebrugh, D. Schumacher, R. J. Wood, and G. C. Wraith, Indexed Categories and Their Applications. VII, 260 pages. 1978.

Vol. 662: Akin, The Metric Theory of Banach Manifolds. XIX, 306 pages. 1978.

Vol. 663: J. F. Berglund, H. D. Junghenn, P. Milnes, Compact Right Topological Semigroups and Generalizations of Almost Periodicity. X, 243 pages. 1978.

Vol. 664: Algebraic and Geometric Topology, Proceedings, 1977. Edited by K. C. Millett. XI, 240 pages. 1978.

Vol. 665: Journées d'Analyse Non Linéaire. Proceedings, 1977. Edité par P. Bénilan et J. Robert. VIII, 256 pages. 1978.

Vol. 666: B. Beauzamy, Espaces d'Interpolation Réels: Topologie et Géometrie. X, 104 pages. 1978.

Vol. 667: J. Gilewicz, Approximants de Padé. XIV, 511 pages. 1978.

Vol. 668: The Structure of Attractors in Dynamical Systems. Proceedings, 1977. Edited by J. C. Martin, N. G. Markley and W. Perrizo. VI, 264 pages. 1978.

Vol. 669: Higher Set Theory. Proceedings, 1977. Edited by G. H. Müller and D. S. Scott. XII, 476 pages. 1978.

Vol. 670: Fonctions de Plusieurs Variables Complexes III, Proceedings, 1977. Edité par F. Norguet. XII, 394 pages. 1978.

Vol. 671: R. T. Smythe and J. C. Wierman, First-Passage Perculation on the Square Lattice. VIII, 196 pages. 1978.

Vol. 672: R. L. Taylor, Stochastic Convergence of Weighted Sums of Random Elements in Linear Spaces. VII, 216 pages. 1978.

Vol. 673: Algebraic Topology, Proceedings 1977. Edited by P. Hoffman, R. Piccinini and D. Sjerve. VI, 278 pages. 1978.

Vol. 674: Z. Fiedorowicz and S. Priddy, Homology of Classical Groups Over Finite Fields and Their Associated Infinite Loop Spaces. VI, 434 pages. 1978.

Vol. 675: J. Galambos and S. Kotz, Characterizations of Probability Distributions. VIII, 169 pages. 1978.

Vol. 676: Differential Geometrical Methods in Mathematical Physics II, Proceedings, 1977. Edited by K. Bleuler, H. R. Petry and A. Reetz. VI, 626 pages. 1978.

Vol. 677: Séminaire Bourbaki, vol. 1976/77, Exposés 489–506. IV, 264 pages. 1978.

Vol. 678: D. Dacunha-Castelle, H. Heyer et B. Roynette. Ecole d'Eté de Probabilités de Saint-Flour. VII-1977. Edité par P. L. Hennequin. IX, 379 pages. 1978.

Vol. 679: Numerical Treatment of Differential Equations in Applications, Proceedings, 1977. Edited by R. Ansorge and W. Törnig. IX, 163 pages. 1978.

Vol. 680: Mathematical Control Theory, Proceedings, 1977. Edited by W. A. Coppel. IX, 257 pages. 1978.

Vol. 681: Séminaire de Théorie du Potentiel Paris, No. 3, Directeurs: M. Brelot, G. Choquet et J. Deny. Rédacteurs: F. Hirsch et G. Mokobodzki. VII, 294 pages. 1978.

Vol. 682: G. D. James, The Representation Theory of the Symmetric Groups. V, 156 pages. 1978.

Vol. 683: Variétés Analytiques Compactes, Proceedings, 1977. Edité par Y. Hervier et A. Hirschowitz. V, 248 pages. 1978.

Vol. 684: E. E. Rosinger, Distributions and Nonlinear Partial Differential Equations. XI, 146 pages. 1978.

Vol. 685: Knot Theory, Proceedings, 1977. Edited by J. C. Hausmann. VII, 311 pages. 1978.

Vol. 686: Combinatorial Mathematics, Proceedings, 1977. Edited by D. A. Holton and J. Seberry. IX, 353 pages. 1978.

Vol. 687: Algebraic Geometry, Proceedings, 1977. Edited by L. D. Olson. V, 244 pages. 1978.

Vol. 688: J. Dydak and J. Segal, Shape Theory. VI, 150 pages. 1978.

Vol. 689: Cabal Seminar 76–77, Proceedings, 1976–77. Edited by A.S. Kechris and Y. N. Moschovakis. V, 282 pages. 1978.

Vol. 690: W. J. J. Rey, Robust Statistical Methods. VI, 128 pages. 1978.

Vol. 691: G. Viennot, Algèbres de Lie Libres et Monoïdes Libres. III, 124 pages. 1978.

Vol. 692: T. Husain and S. M. Khaleelulla, Barrelledness in Topological and Ordered Vector Spaces. IX, 258 pages. 1978.

Vol. 693: Hilbert Space Operators, Proceedings, 1977. Edited by J. M. Bachar Jr. and D. W. Hadwin. VIII, 184 pages. 1978.

Vol. 694: Séminaire Pierre Lelong – Henri Skoda (Analyse) Année 1976/77. VII, 334 pages. 1978.

Vol. 695: Measure Theory Applications to Stochastic Analysis, Proceedings, 1977. Edited by G. Kallianpur and D. Kölzow. XII, 261 pages. 1978.

Vol. 696: P. J. Feinsilver, Special Functions, Probability Semigroups, and Hamiltonian Flows. VI, 112 pages. 1978.

Vol. 697: Topics in Algebra, Proceedings, 1978. Edited by M. F. Newman. XI, 229 pages. 1978.

Vol. 698: E. Grosswald, Bessel Polynomials. XIV, 182 pages. 1978.

Vol. 699: R. E. Greene and H.-H. Wu, Function Theory on Manifolds Which Possess a Pole. III, 215 pages. 1979.